The Travels of Jedediah Smith

". . . yet was he modest, never obtrusive, charitable. . . . a man whom none could approach without respect, or know without esteem. And though he fell under the spears of the savages, and his body has glutted the prairie wolf, and none can tell where his bones are bleaching, he must not be forgotten." From **Jedediah Strong Smith,** an anonymous eulogy in **Illinois Monthly Magazine,** 1832. The above sketch of Jedediah Smith was said to have been made by a friend, from memory, after Smith's death.

The Travels of Jedediah Smith

A DOCUMENTARY OUTLINE
Including the Journal of the Great American Pathfinder

By
MAURICE S. SULLIVAN

University of Nebraska Press
Lincoln and London

First Bison Book Printing: 1992
Most recent printing indicated by the last digit below:
10 9 8 7 6 5 4 3 2 1
Library of Congress Cataloging-in-Publication Data
Smith, Jedediah Strong, 1799–1831.
The travels of Jedediah Smith: a documentary outline including the journal of the great American pathfinder / [edited] by Maurice S. Sullivan.
p. cm.
Originally published: Santa Ana, Calif.: Fine Arts Press, 1934.
"Bison book"—T.p. verso.
ISBN 0-8032-4219-0 (cloth).—ISBN 0-8032-9206-6 (pbk.)
1. West (U.S.)—Description and travel—To 1848. 2. Overland journeys to the Pacific. 3. Fur trade—West (U.S.)—History—19th century. 4. Smith, Jedediah Strong, 1799–1831—Journeys—West (U.S.) I. Sullivan, Maurice S., 1893–1935. II. Title.
F592.S647 1992
917.804′2—dc20 91-46728
CIP

Reprinted by arrangement with Theresa Happe

The title of the original edition, published in 1934 by the Fine Arts Press, Santa Ana, California, was *The Travels of Jedediah Smith: A Documentary Outline, Including the Journal of the Great American Pathfinder*. The 1839 map by David H. Burr, probably based on a sketch or notes by Jedediah Smith, is omitted from this Bison Book edition because it could not be reproduced legibly.

∞

List of Illustrations

Foreword

FIFTEEN years ago, if the average university student or the average educated man had been asked what he knew about Jedediah Smith, this average person would have been obliged to admit that he had never heard even the name of the great American pathfinder.

He could have told something of Frémont, the official explorer, who was a child when Jedediah Smith, trapping and trading, already had examined Western America from Mexico to Canada, and from the Missouri River to the Pacific Ocean.

He could have told something of Kit Carson, the worthily celebrated scout, who, though he appeared in the Far West after it was familiar to Jedediah Smith, was fortunate enough to find a biographer, and thus to become famous. A few educated Americans were aware of Bonneville, basking in the reflected glory of Washington Irving.

The inquirer might have searched encyclopedia and school book without learning the name of the discoverer of the central route from the Rocky Mountains to the Pacific Ocean; the man who made the effective discovery of the South Pass; the first white man to cross Nevada; the first to traverse Utah from north to south, and from west to east; the first American to enter California by the overland route; the first white man to conquer the High Sierra; the first to explore the entire Pacific Slope from San Diego to Vancouver; in brief, the man who charted the way for the spread of the American empire from the Missouri River to the Western Sea.

In the course of the last fifteen years, and especially since 1926, the hundredth anniversary of his entry into California, the name of Jedediah Smith has become better known.

Two things, chiefly, have contributed to the amazing neglect of this extraordinary person.

The first contributor was death. The Indians killed him while he was still a young man.

The second was fire: fire in Missouri, fire in Illinois, fire in Kansas, fire in Iowa, fire in California, fire in New Mexico and fire in Canada.

Jedediah Smith had kept daily records of his eight years of adventure in mountain and desert; in places, some of which no white man had ever trod before. He intended to publish his diaries, and a map which changed the geographic conception of North America west of the Rocky Mountains.

Specialists in Western Americana, curators of historical societies and other interested persons began a search for these journals a generation ago.

Everywhere, among those likely to know, it was said the journals had been burned. Some said they were burned in the warehouse of Tracy and Wahrendorff, at St. Louis. Others said they were lost in a dwelling fire while in the keeping of Washington Irving, who, it was reported, had intended to do for Smith what he had done for the Astorians and Captain Bonneville. Still others affirmed that the journals were destroyed in a fire at the home of a St. Louis writer.

The indefinite nature of these rumors, whether in print or tradition, and the possibility that here and there throughout the country, in the bottoms of old trunks, might be found letters and other papers which would reveal much about the travels and character of a neglected hero, led the author of this volume to undertake tracing the descendants of the numerous brothers and sisters of Jedediah and of his relatives by marriage in the Simons family. The task required several years. The members of the living generation were scattered from Florida to Washington and from Canada to the border of Mexico, some unaware of the existence of the others.

In the end it became virtually certain that the original journals had indeed been burned, long after the death of Jedediah Smith and under circumstances which need not be recorded. At other places and other times other Smith documents were accidentally lost.

However, among the papers of an isolated branch of the Smith family, the members of which were well aware of the greatness of their kinsman, a verbatim copy of some of the most important material in Jedediah's notes happily has been preserved. The copy was made by Samuel Parkman, a friend of the Smith family. He was the "Mr. S. Parkman" mentioned in the Eulogy of Jedediah Smith, published in the *Illinois Monthly Magazine* in 1832; he should not be confused with the Rev. Samuel Parker, or Francis Parkman, both writers of history.

This was the copy which Jedediah intended to publish. It is listed in the inventory of his estate:

> *A Manuscript Journal of the travels of Jedediah S. Smith thro' the Rocky Mountains and West of the Same together with a description of the Country and the Customs and Manners of the different Tribes of Indians thro' which he travelled.*

The *Journal* is a record of his advent into the fur trade, of his remarkable walk across the Utah desert, of his second journey into California before any other American had penetrated there overland, of his escape from the Mojaves, and of his adventures in California and Oregon.

This valuable manuscript, never before available to students of history, is presented in the following pages. For the first time the exact routes of the southwestern, trans-Utah, California and Oregon travels may become known, in addition to Smith's own account of the Indians he met, many of whom never saw a white man until they laid eyes on him.

Besides the *Journal* there is a map, drawn in 1839. This map, entirely overlooked by scholars who have contradicted one another about the routes of Jedediah Smith, plainly is based on Jedediah's own directions, if not actually copied from his long-missing chart. On it the trails blazed by Smith are definitely shown.

Supplementing the Smith *Journal* are the diary of Alexander Roderick McLeod, describing events on the Hudson's Bay Company expedition to recover property of Jedediah Smith, carried off by the Indians after the massacre on the Umpqua River, Oregon, and pertinent correspondence. These documents are published with the generous permission of the Governor and Committee of the Hudson's Bay Company, London.

I wish here to acknowledge the aid given me, especially by the late Walter Bacon, the late Mrs. Kate S. Etter, J. Ernest Smith and George W. Beattie.

Mr. Bacon, son-in-law of Peter Smith, permitted access to numerous enlightening papers of the Smith family, and from the beginning took a lively interest in this work. Mrs. Etter and her brother, J. Ernest Smith, the best-informed living relative of Jedediah Smith, his grandvncle, gave me permission to print the Smith *Journal,* and readily provided a wealth of dependable material based on conversation with several of his immediate family. Mr. Beattie, who first interested me in the story of the great American pathfinder, in the course of several years assisted not only in allowing the use of his collection of books of reference and documentary transcripts, but also by invaluable counsel and encouragement.

To Miss Laura C. Cooley, of the excellent history department of the Los Angeles Public Library, and to Miss Edith DeMond, reference librarian of the A. K. Smiley Library, Redlands, California, I am indebted for frequent use of rare printed sources, and to Dr. Herbert I. Priestley, librarian of the Bancroft Library, University of California, for copies of important documents. Besides, the Library of Congress, the Warner Library, Pennsylvania State Library, the California State Library and public libraries at Riverside, Santa Rosa, Pasadena, and Oakland, (Cal.), Boston and New Bedford, (Mass.), St. Louis and St. Joseph, (Mo.), Jacksonville and Mount Vernon, (Ill.), Akron, Mount Vernon, and Mansfield, (O.), Council Bluffs, (Ia.), Allentown, (Penn.), Great Falls, (Mont.), and Binghamton, (N. Y.) courteously and carefully answered inquiries.

A twentieth-century book on the American fur trade and Western exploration may have been written without the aid of the manuscript collection of the Missouri Historical Society, but if so it was an exception. Miss Stella Drumm, librarian, and Miss Anne Kinnaird, assistant archivist, have been very helpful. So also Kirke Mechem, secretary of the Kansas Historical Society. Lewis A. McArthur and Miss Nellie B. Pipes of the Oregon Historical Society assisted with information on Oregon place names. In addition I have called upon the resources of California, Illinois, South Dakota, Wisconsin, Montana, Washington, Oregon, Idaho, Western Pennsylvania and Marine Research (Salem) Societies.

Ralph T. Smith, Mrs. C. F. Calhoun, Mrs. J. M. Green, Mrs. C. C. Bentley, J. Smith Bacon, F. R. Bacon, Mrs. Lurah F. Jenkins, Mrs. R. D. Baker, Mrs. Albert Bondurant, Mrs. Florence Skang, Wilson S. Dalton, Mrs. W. F. Familton, Mrs. Victor P. Gay, Mrs. R. P. Gannon, Mrs. James Gray, David D. Roberts, Mrs. Julia H. Tyler, Guy S. Carleton, Miss May Davis, Mrs. Edna Smith Jeffrey, Mrs. Ivan Messenger, and W. B. Fox, all members of the Smith or Simons families, or of both, have done what they could to assist. The editors of the Ashtabula (O.) Star-Beacon, Kendallville (Ind.) News-Sun, Plymouth (Ind.) Pilot, and Lewis (Ia.) Standard, Benjamin F. Boydstun, Waco, (Tex.), the Rev. M. C. Wright, Kendallville, Miss Mae G. Foster, of the Society of California Pioneers, Mrs. Mary Thompson Deady, San Francisco, and Mrs. Elizabeth C. Renner, Simons, (O.), also gave assistance in tracing Smith and Simons descendants.

For photostats of supplementary documents I am indebted to Miss Anne M. Southwick, of St. Louis, and for special information to Miss Adele Ogden, Berkeley; Miss Clara Parker, Nantucket; Ralph W. Kirby, Bainbridge; the Rev. Fr. Zephyrin Engelhardt, Santa Barbara; State Engineer Edward Hyatt, Sacramento; Assistant State Engineer Max F. Rogers, Salem (Ore.); Major A. M. Barton, Sacramento; Dr. A. M. Woodbury, Salt Lake City; Charles Kelly, Salt Lake City; Dr. M. R. Harrington, Los Angeles; H. L. Reid, Zion National Park, County Surveyor Frank E. Kelly, Eureka, (Cal.); Dr. N. D. Mereness, Washington, D. C.; Henry R. Wagner, San Marino, (Cal.); F. M. Kelly, Needles, (Cal.); Dr. C. Hart Merriam, Washington, D. C.; Mrs. Alice B. Maloney, Santa Ana, (Cal.); and the United States War Department.

The *Secretaria de Relaciones Exteriores, Departamento de Archivos y Servicios Interiores,* Mexico, D. F. kindly sent me a copy of a letter of Austin Smith.

Finally, I must acknowledge the usefulness of the comprehensive bibliography prepared for the Historical Society of Southern California by Dr. A. P. Nasatir, San Diego, and both the bibliography and references of Harrison Clifford Dale's *Ashley-Smith Explorations.*

This is not a biography of Jedediah Smith. The preparation of that is a work upon which the writer is engaged, with the object of presenting for the man in the street the complete story of the pathfinder. In this volume is a basic outline of the career of Smith, beginning with his entry into the fur trade at the age of twenty-three. Out of the quest has come other century-old material with much intimate information about the pathfinder, not included here.

What charm there is for the enthusiast in finding a letter beginning "An Unworthy Son Once More Undertakes to Address his Mutch Slighted Parents"! Or the story of Jedediah's pet beaver with the red collar—the beaver which suddenly answered the call of the wild and which, if apprehended, would have brought to its captors princely rewards in coffee and sugar and rum. Or in looking upon the stained claw of the very grizzly bear that tore Jedediah's scalp and one of his ears from his head so that they must be sewed in place with household needle and thread. Or in seeing the marvelously complicated watch of which the Comanches robbed him and which found its way to Santa Fé in the hands of *comancheros,* to the California gold diggings in the hands of a friend, and finally, long years afterward, back into the possession of the kinsmen of the owner. Or in opening a drawer in a little mirrored dressing case carried in the wilderness and there coming upon Jedediah's snuff box. And what satisfaction there is examining intimate letters and records and finding no slightest reason to doubt that Jedediah Smith was other than he seemed to be: a man apart when civilized men turned barbarian, comforting themselves with the saying that "God holds no man accountable after he crosses the Missouri River."

Little by little, Jedediah Smith, a giant of the Western world, emerges from a century of unconcern. James Clyman breaks through the veil of obscurity and tells of the young 'Diah, lying torn and bleeding on the ground, directing the crude surgery of the mountain men. Out from attics come papers falling apart at the creases, contributing their mite toward illuminating the personality of the pathfinder.

Just as the *Journal* of Jedediah Smith, the diary of his protégé, James Clyman, the correspondence of Sir George Simpson, and other documents are disclosed a century after the events recorded therein, so, doubtless, more enlightening papers will appear in the future.

In order that the frontier background of Jedediah Smith might be presented as it was, the original spelling and punctuation of the documentary material has been retained wherever possible. Neither Jedediah Smith nor the Smith family nor Samuel Parkman (who probably altered Jedediah's spelling) need apologize for orthographic imperfection shared with Washington, Jefferson, Lewis and other early Americans; men educated in the best sense of that word. There was in the letters of the pioneer Smiths and the related Strongs a natural dignity of expression, which a misspelled word could not mar.

The purpose of this book is to provide a sound basis of fact as a new starting point; if anything else, whether great or small, has remained forgotten, or if anyone can throw new light on the Smith trails, the editor would be glad to have it called to his attention in a note addressed to 8 Beacon Street, Redlands, California.

Here then is an outline of the story of the pious, but none the less vigorous, fighting Knight in Buckskin, the man who opened the gate through which passed the American builders of the West. It begins with a fragment of his own narrative.

MAURICE S. SULLIVAN

The Travels of Jedediah Smith

I HAD passed the summer and fall of 1821 in the northern part of Illinois, and the winter of 21 & 22 at or near the Rock River rapids of the Mississippi.[1] In the spring I came down to St. Louis and hearing of an expedition that was fiting out for the prosecution of the fur trade on the head of the Missouri, by Genl Wm H. Ashley[2] and Major [Andrew] Henry,[3] I called on Genl Ashley to make an engagement to go with him as a hunter. I found no difficulty in making a bargain on as good terms as I had reason to expect. On the 8th of May I left St. Louis on board the *enterprize* under the direction of Daniel S. D. More.[4] But before I proceed further it may be necessary to give a brief outline of the business commenced by Genl Ashley and Maj. Henry. . . . .[5]

Leaving St. Louis our boat proceeded on without any material occurrence for the first three hundred miles. The strong current of the Missouri made the voyage slow, Laborious and dangerous. Arrived at a place within the State of Missouri and near the mouth of the Sni Eber Creek[6] on a windy day and turning a point full of sawyers the boat by an unexpected turn brought the top of her Mast against a tree that hung over the water and wheeling with the side to the powerful current was swept under in a moment. The boat and its valuable cargo worth [10,000] Dollars was lost with the exception of a few articles that floated and were saved by the exertion of two or three active men.

After the loss of the boat Mr. More immediately started for St. Louis, Leaving the party, and myself among the rest, at or near the place where it was lost. About the 4th of June Mr. More arrived in St. Louis and gave Gen[l] Ashley intelligence of the loss of his boat. Not discouraged by this unfortunate occurrence, Gen[l] Ashley immediately commenced fitting out another boat and in Eighteen days was prepared to Leave with another boat and cargo and 46 men. He then took charge of the Expedition himself and procided up to the place where we were encamped without any verry material occurrence.

The Country from the place where Gen[l] Ashley Joined us for a long distance up the Missouri has been frequently and well described. I shall therefore pass it over and merely make an occasional remark on those things that struck me most forcibly at the time. Gen[l] Ashley, contrary to the common custom of traders of this river, had laid in a plentiful supply of provision, consisting of sea Bread and Bacon, so that we were not dependant on the precarious supply derived from hunting, although at the same time that the boat was moving against the powerful current a few men who were good and active hunters were out on the bank hunting for such game as the country afforded, which consisted of Black Bear, Deer, Elk, Raccoon and Turkeys in abundance. And as the Country was well stocked with Bees we frequently had a plentiful supply of honey. For some distance up the Missouri the country is verry fine, and as the Gen[l] kept me constantly hunting, to which I was by no means averse, I was enabled to enjoy the full novelty of the scene in which I was placed and at the same time avoid the dull monotony of following along the bank of the river entirely dependent on the motions of the boat.

In the progress of the Journey we passed the mouth of the River Platte, and about [40] miles above the Platte was The United States Post called Council Bluffs.[7] At that time there was something like [490] troops stationed at that place under the command of [Lieut.-Col. Henry Leavenworth].

From the Council Bluffs continuing our voyage up the river we passed the villages of the Mahaw's, Puncah's[8] and some bands of the Sioux and arrived in the heart of the Sioux. The indians came in to see us frequently and the Gen[l] held council with them and smoked the pipe of Peace. These visits gave me frequent opportunities for observing their manners, but as travelers have often described them I will say nothing on that subject at present.

The Puncah's and the Mahaws live in stationary villages and raise corn, but the Sioux at that time had no permanent residence, but moved about wherever the game was most abundant. The Range of the Sioux extended from the foot of the Black hills across the Missouri to the River St. Peters and the Mississippi, and along the Missouri from the villages of the Puncahs to the Arichara[9] towns.

This country may be considered one extensive prairae interrupted only by the narrow fringes of timber along the rivers, the surface gently undulating and covered with grass. No mountains. In some parts immense herds of Buffalo. Antelope in abundance. Some Deer, Bear and Elk and some Deer of the Blk tail kind.

The Lodges of the Sioux are made of Skins, the frame being of light poles from 12 to 16 feet in length, being set on the ground in such a manner as to make the Lodge present when the Skins are put on the frame, the appearance of a stack of grain or perhaps a Large Shock of tall hemp, with exception that the Lodge has generally from eight to ten sides. The diameter of the bore of a common sized Lodge is perhaps from 10 to 12 feet and the heighth about 18 feet. The taper from the bore is regular to the top, where the poles are fastened to gether and where there is a small aperture left for the smoke. A small hole is left at the side for a door over which a skin is hung, which is raised when passing in or out and falls back to its place without any assistance.

Going into the interior of the Lodge you find the fire built in the center and the furniture, principally of Skins, kettles, axes, arms, meat, &c spread around the circular room, leaving a small vacant space around the fire. These Lodges are verry comfortable, being in winter much warmer than would be supposed. The Skins of which the Lodge is composed are thick and the seams by which they are attached together are so close that [while there is verry little opportunity for a][9a] they present a complete defence against the wind. In the winter they pile Blocks of turf on the Lower edge of the skins to keep them close to the ground, and the Skins are also fastened by stakes driven in the ground through loop holes for that purpose.

A small fire is sufficient to make them quite warm in the coldest weather. They do not smoke except from a sudden change of wind and then no longer than it takes a squaw to spread a smoke wing of the Lodge skin, which is calculated for that purpose, to the windward of the aperture at the top. In summer the cooking is all done outside and the Lodge is made quite cool by raising the lower edge of the skin and leaving an aperture all around the circle of about two feet in width for the passage of the air.

The distant appearance of [one][9b] of these [encampments was][9c] Lodges when many indians are encamped to gether cannot fail of pleasing. Clustered to gether with their yellow sides and pointed tops, the children playing around in the intervals between them, the men going out or coming in from hunting, The horses feeding on the neighboring prairae, the dogs (of them are great numbers) sleeping or playing in the sun or shade, the squaws at their several Labors and the boys at their several Sports. These, taken in connection with a beautiful mingling of prairae and woodland or some undulation of the land or some bend of the great River that brings them at once to view, and above all eyes that are not accustomed to such a sight, would almost persuade a man to renounce the world, take the lodge and live the careless, Lazy life of an indian.

Some bans of the Sioux are rich in horses, but many of them do their packing principally on dogs.

The Sioux sometimes paint their Lodges with the figures of Buffalo, Deer, horses, Battles and many other things which they consider interesting, which gives to the encampment a gaudy and verigated appearance. On [entering][9d] approaching a Lodge you observe at the door three straight and handsome poles set up in a triangular form and joined to gether at the top, on which is suspended the Medicine sack of the owner, consisting of such things as he fancies to possess a certain undefined charm. Over these is hung perhaps a piece of scarlet or a red Blanket and the Skin of a white wolf or something which is highly valued. [Going up to][9e] Entering the Lodge, on the opposite side from the door, you observe the Master of the Lodge sitting on a skin or Robe and leaning back with no borrowed dignaty against a mat formed of willows pealed and some other material which answers the purpose of keeping them together, which is supported by three sticks set up in a triangular form, the mat leaning against them and supported in a sloping position by two of them. The women, as they have all the work to perform, sit next the door and also have their tools in the same part of the lodge that they may be employed about their several employments without disturbing the men.

The Sioux are generally above the common stature and of a complexion somewhat lighter than most indians. They have intelligent countenances and are in person generally good looking men. In the moral scale, as their appearance would indicate, they rank above the mass of indians.

Continuing our journey we arrived at the Arickara towns on the 8th of Sept. There are two villages situated about three hundreds yards from the Missouri and on the left bank. They are about a quarter of a mile apart and consist of about Lodges in each village.

As these indians are not roving bands like the Sioux they build their lodges in a more permanent manner. They are circular in form and composed of forks set in the earth to support the poles and split timbers with which they are covered. From the sides they rise to form the roof by a gentle slope to the highest point, which is the center of the building, to which place all the timbers of the roof [extend][9f] converge, and are supported by a fork or Post longer and Larger than those at the sides. The roof having been prepared to sustain a considerable weight is first covered with grass or some light material and is then covered to a considerable debth with earth. The sides are also embanked with the same material and the house when finished presents the appearance of a regular mound of earth. The interior arrangement is like that of a lodge with the exception that a narrow scaffold is built around next the side of the building on which they sleep. Some of these Lodges have a covered entrance made of the same material as the body of the Lodge, about 4 feet wide and ten feet long. The country in the vicinity of the Arickara villages is much like that spoken of in my remarks on the Sioux country.

Immediately on the arriving of the boat at the Arrickaras Gen[l] Ashley determined, as the season was much advanced, to purchase horses and proceed directly by land to the Mouth of the Yellow Stone whilst the boat, which would proceed more slowly, would continue on to the same place.

The Gen[l] took charge of the party that went by land himself and to this party I was attached. He moved with great care, being somewhat apprehensive of danger from the Arickara indians.

In our route this far we had seen and killed some Buffalo, but they were not in great numbers, but on the evening of the second day after leaving the Arickara's it seemed to my unaccustomed eyes that all the buffalo in the world were running in those plains, for far as the eye could see the plains and hills appeared a moving body of life. Over the hills and plains they moved in deep, dense and dark bodies resembling the idea I have formed of the heavy columns of a great army. As they took the wind of the party they ran, making the ground tremble with the moving weight of animal life.

Perhaps since that time I have frequently seen as many or more Buffalo than were in view at that time, but they never made that strong impression that was made by the ten thousands of that day that seemed sufficiently numerous to eat everything like vegetation from the face of the country in a single week.

Still in our progress the Buffalo were seen in numbers apparently inexhaustable and innumerable. The Country generally a prairae gently undulating and well calculated to support the immense drafts that were made on the vegitation by the numerous animals that claimed it as their pasture ground.

In about   Days the party arrived at the Mandan villages. The Gen[l] and his party were invited into a large Lodge and some of the principal chiefs being invited, the pipe of peace was smoked according to the usual ceremonies. A council was held and the Gen[l] said such things to them as he thought most likely to secure and continue their friendship.

The Mandans and the Gros Ventres live like the Arickaras in fixed villages. All the villages of these two tribes have been frequently described. I will there fore pass by them and continue on our journey. We left the Mandans after remaining but one day and pushed on towards the Yellow Stone.

In the progress of this journey we crossed the Little Missouri, remarkable even in comparison with the Large river for its muddy, turbid waters. This river and the White Earth river on the other side of the Missouri contribute a great proportion of the mud of the Missouri River below their Mouths.

The country from the Mandans had been of the same character with that before described. The surface gently undulating and (near the Little Missouri somewhat Broken) watered by Creeks and Springs. Well clothed with grass and one of the finest Buffalo Countries in the world. From the Little Missouri the country continues much the same as that before described, except that it is not as well watered. There are some streams in the interior, but during the dry season the Buffalo generally resort to the larger River for water.

The Party arrived at the mouth of the Yellow Stone on the 1st Day of October. Gen[l] Ashley and Maj. Henry immediately commenced arrangements for business, and after furnishing the mountain parties with their supplies of goods and receiving the furs of the last hunt Gen[l] Ashley started for St. Louis with a large Pirogue   Packs of Beaver and Men.

After the departure of Gen[l] Ashley for St. Louis the company of hunters under the direction of Maj. Henry were busily engaged in making preparations for trapping. Mr. Chapman and myself went with a small party of men up the Yellow Stone a short distance for the purpose of procuring a supply of meat for the fort[10] and such skins as were wanted for the use of the company, in the mean time to take what Beaver we could conveniently.

In the mean time a party had gone up the Missouri in a Boat and canoes for the purpose of trapping under the immediate charge of Maj. Henry, and another in canoes up the Yellow Stone for the same purpose. When Maj. Henry left the fort it was his intention to ascend the Missouri as far as the mouth of Milk River and the party on the Yellow Stone were by instruction to ascend that river [as far as][10a] to the mouth of Powder River and up it as far as practicable.

After we had finished the business on which we were engaged we returned to the fort. Soon after our return Mr. Chapman and myself with a party of [?] men, whose names are in the margin, ascended the Missouri, traveling immediately along the bank of the river. We were equiped as far as we knew at the time in good order for trapping and hunting.[11]

On our way up we met Maj. Henry [near][11a] on his return, and after making the necessary arrangements we continued our journey up the river. The country has been well described by Lewis & Clark, therefore any observations from me would be superflous.

During the principal part of the journey we found a plenty of Buffalo, although that was not the season in which they frequented that part of their range. About the first of November we arrived at the mouth of the Muscle Shell River, which was the place of our destination. The River was fast filling with ice and we were admonished to prepare for the approaching winter, which in that latitude must of course be much colder than the winters of the country from which most of us came.

We were generally good hunters, but at that time unacquainted with the habits of the Buffalo, and seeing none in the vicinity we supposed they had abandoned the country for the winter. We therefore became somewhat apprehensive that we should suffer for want of provisions. While a Part of the company were engaged in preparing houses for the winter, I took some of the best hunters and made every exertion to procure a supply of meat sufficient for our suport. And we were indeed verry successful, for we killed all the small game of the vicinity, particularly antelope and deer, Laying up a supply of meat that drove the apprehension of want entirely from our minds.

Our houses being finished, we were well prepared for the increasing cold. When the weather had at length become extremely cold and the ice strong and firm across the River, we were astonished to see the buffalo come pouring from all sides into the valley of the Missouri, and particularly the vast Bands that came from the north and crossed over to the south side on the ice. We there fore had them in thousands around us and nothing more required of us than to select and kill the best for our use whenever we might choose.

The place we had selected for winter quarters was about [?] Miles [?] the mouth of the Muscle Shell River in a situation well calculated for the supply of such things as were necessary for our safety and comfort, particularly a great sufficiency of Cotton wood timber, an article indispensibly necessary in wintering horses. In feeding the horses the trees are cut down and the bark is shaved off and given to them in any quantity they can consume. The small limbs are placed before them and they soon learn to eat off the bark without the trouble of Shaving or rinding. In shaving the Bark from the larger trees the outside rind is thrown away. Forced to eat this bark from necessity, though perhaps not much averse to its taste, horses soon become verry fond of it and require little other assistance than the felling of the trees, and, strange as it may appear to those unacquainted with such things, they become fat and will keep so during the winter if not used. But the fat acquired by this food is not permanent and is worn off by a few days of hard riding.

In our little encampment, shut out from those enjoyments most valued by the world, we were as happy as we could be made by leisure and opportunity for unlimited [gratification of the][11b] indulgence in the pleasure of the Buffalo hunt and the several kinds of sport which the severity of the winter could not debar us from.[12]

* * *

"Mr. Smith, a young man of our company, made a powerful prayer, which moved us all greatly, and I am persuaded John died in peace." Hugh Glass, writing to the parents of John Gardner, wounded in the battle with the Aricaras, in 1823. Painting, by Charles Holloway, in State Capitol, Pierre, S. D.

At St. Louis, early in the year 1823, General Ashley recruited another large troop of hunters, trappers, boatmen and servants. This time he was obliged to accept men whom he would have rejected when he organized the first expedition.

James Clyman,[13] one of the volunteers, compared the company to the rapscallion regiment of Falstaff. There were, however, young men of worth in that motley crew: Clyman himself, Thomas Fitzpatrick[14] and William L. Sublette,[15] for examples.

As the expedition proceeded up the Missouri River, some of the less desirable deserted, and from time to time new and better recruits were gathered. All went well as far as the Aricara villages. A short party, guarding horses obtained from the Rees by General Ashley, was fired upon.

Trapped between the Ree stockade and the river, and at first unable to get help from the boatmen, the whites fought courageously until retreat was possible.

Here we get our next view of the young 'Diah Smith, whose *Journal* left him winter-bound near the mouth of the Musselshell River, in what is now the State of Montana. Where he joined the second expedition is not apparent in available records. Probably he was the messenger sent by Major Henry, speeding news of a battle with the Blackfeet, and of the urgent need for horses to replace some stolen by the Assiniboines.

Jedediah was in the shore party at the Ree villages. Only twenty-four years of age, he was now an experienced mountain man, one of the *hivernans;* a hunter who had wintered in the Rocky Mountains. He took charge of the hopeless defense of the shore party until aid came at length from the boats.

General Ashley lost twelve men in that engagement. Twelve others were wounded, of whom one died.

The party dropped down the river to await reinforcement to punish the Rees. Jedediah Smith, who was familiar with the overland route to Henry's post, volunteered to go to the

Yellowstone and get as many of the major's men as he could spare. Smith made the round trip in less than a month. His own record of the journey has been lost, but his feat was the subject of many tales repeated by mountain men at the campfires of the wilderness and in the taverns of St. Louis.

Major Henry's canoes, carrying the winter cache of furs, slipped by the Ree villages in the night. Jedediah Smith took the peltry down to St. Louis, reported the Ree attack to the military authorities, and hurried back to rejoin General Ashley.

Lieutenant-Colonel Henry Leavenworth with 250 soldiers, Joshua Pilcher of the Missouri Fur Company with sixty mountain men, and a band of the Aricaras' enemies, the Sioux, advanced upon the Ree villages. Smith was captain of one of the two companies of Ashley-Henry men.

After several days of gun-fire, stratagem and parley, all confused by Leavenworth's preference for peace rather than war, the Rees retreated under cover of night. Later the military commander testified that he counted thirty-one new graves in the Aricara villages.[16]

General Ashley returned to St. Louis to raise more money. Major Henry struck out for the Yellowstone by land, taking the majority of the Ashley-Henry men with him.

A smaller expedition was placed in charge of Jedediah Smith. The party went westward from Fort Kiowa, southeast of the present city of Pierre, South Dakota, After passing the Black Hills, and near one of the forks of the Cheyenne River, the cavalcade was surprised by a grizzly bear, which leaped upon Smith, knocked him down, bit and clawed him before it was killed.

One of Jedediah's ears was torn almost from his head; his scalp was horribly gashed. Clyman records that Smith, lying on the ground, directed the treatment which resulted in his recovery, ordering Clyman to sew the wounds on his scalp with household needle and thread, and to stitch his ear in place.[17] Despite this injury, Smith was able to ride his horse into camp.

Ten days later he was again leading the party. He trapped the Powder River and neighboring streams in what are now Montana and Wyoming. The fall hunt over, the group wintered among the Crow Indians at an encampment on the Wind River.

Although the first westbound crossing of the famous South Pass by a white man has been attributed to Thomas Fitzpatrick, Clyman's statement makes it apparent that Smith, who was commander of the company, late in February of 1824 led his men up the Sweetwater and in March arrived on the Big Sandy River, across the Continental Divide.[18]

A group of Astorians, returning from Oregon, appears to have taken that trail in 1812.[19] Perhaps some private in the shadowy company of forgotten men went westward through the South Pass before 1824; but to Jedediah Smith, if Clyman's memory is not at fault, must go the credit of opening the route as the great highway to the Far West.

The Ashley men were divided into small parties under the leadership of Smith, Fitzpatrick and Clyman, who took beaver from the tributaries of the Green River in a spring hunt. In the summer they returned to rendezvous on the Sweetwater, whence their furs were sent down to St. Louis.

Major Henry retired, and young Smith became his successor. He then began that series of explorations which made him the pathfinder of American pathfinders.

His first long journey combined trapping with an investigation of British fur trading activity. He and a party of six men encountered Alexander Ross and a Hudson's Bay Company expedition, and went with them to see what there was to see, traveling northward to the Hudson's Bay post among the Flathead Indians in the northwestern part of what is now

Montana.[20] From Flathead Lake Smith penetrated as far as as the present Canadian line.

In December Smith met Peter Skene Ogden, another Hudson's Bay trader, and with him began the return journey.

"He fell on the waters of the Grand [Salt] Lake," General Ashley wrote to General Atkinson. "He describes the country in that direction as admitting of free and easy passage and abounding in salt. At one place particularly hundreds of bushels might have been collected from the surface of the earth within a small space. He gave me some specimens, which equal in appearance and quality the best Liverpool salt."[21]

Lacking Smith's own story of the journey, we cannot tell whether this was his first visit to the Salt Lake. On the basis of the only evidence available, the honor must be given either to Étienne Provot or James Bridger.

In 1825 General Ashley explored the Green River. In the summer he returned to the rendezvous, gathered the pelts "collected at the expense of severe toil, privation, suffering, peril, and, in some cases, loss of life,"[22] and accompanied by Jedediah Smith and a large party of his men transported them by way of the Big Horn, Yellowstone and Missouri Rivers to St. Louis.

At the summer rendezvous of 1826 Ashley sold his business to Jedediah Smith, David E. Jackson[23] and William L. Sublette, and returned to St. Louis with the understanding that he should send pack trains to the mountains with supplies and trade goods for the new firm.

To Jedediah Smith the partners allotted the task of exploring the country to the southwest, locating new trapping grounds, and, as it proved, opening the way for the advance of the American people from the Missouri River to the Pacific Ocean.[24]

Late in August of 1826, while Jackson and Sublette set out to trap the mountain country, Smith and seventeen[25] mounted men left the Great Salt Lake and rode southward to the Sevier

River, pausing on the way to make a treaty with the warlike Utes.

Smith took with him led animals loaded with knives, awls, tacks, mirrors, arrow points, balls, powder, razors, rings, bells and other articles coveted by Indians. Wherever the natives came to him he gave them presents and tried to establish friendly relations for the future.

He passed up the Sevier River on the east side, crossed over above the mouth of Clear Creek, followed that stream and then struck south, fording a river which he called Lost[26]—the one now known as the Beaver. He came at length upon the Virgin River, to which he gave the name Adams, in honor of the President of the United States.

By now the party was destitute of provisions, and Smith, as he recorded, "had learned what it was to do without food."[27] Starved, thirsting horses sank exhausted, and were despatched for the sake of their poor meat.

The pathfinder followed the Virgin River with difficulty, until he came to the Colorado River. He crossed the Colorado to the east side and went down to the villages of the Mojave Indians, with whom he traded amicably for provisions.

Among the Mojaves were some Indian deserters from the Spanish missions, who had in their possession horses stolen from the mission lands. Smith now had left fewer than half the fifty horses with which he started his journey; the survivors were exchanged for fresh mounts, and others purchased. Here also he obtained information of the Spanish settlements, and with two Indian guides set out across the Mojave Desert, followed the Mojave River—which he named Inconstant—to its source, crossed the San Bernardino Mountains, and, late in November, 1826, traveled to Mission San Gabriel, near Los Angeles.[28]

Jedediah Smith thus completed the first recorded journey overland from the Missouri River to California.[29]

Father José Bernardo Sánchez, head of Mission San Gabriel, gave the company the greatest of hospitality. So kind was Father Joseph, as Smith called him, that Harrison Rogers, clerk of the expedition, exclaimed in his diary:

> Old Father Sanchus has been the greatest friend that I ever met with in all my travels, he is worthy of being called a Christian as he possesses charity in the highest degree—and a friend to the poor and distressed, I ever shall hold him as a man of God, taking us when in distress feeding and clothing us—and may god prosper him and all such men.[30]

This despite the fact that Rogers looked with disapproving eye upon the prevailing theology. Later, when deeds unbecoming to visiting tourists were alleged against the Americans—laying claim to San Joaquin Valley, inducing Indian neophytes to desert the missions, fraternizing with enemies of the Christian Indians, and neglecting to leave California in compliance with the Governor's order—they had worn out their welcome, and found the missionaries in the north not so glad to see them as Father Sánchez had been.

From San Gabriel Jedediah Smith rode down to San Diego to meet Governor José María de Echeandia,[31] by whom he was received with polite suspicion. The Governor refused to allow the American company to travel up the coast to Bodega, a Russian settlement northwest of San Francisco, but after long parleying,[32] and when at length Smith convinced him that he was merely a hunter and not a captain of soldiers,[33] Echeandia gave the party a passport to leave the settlements on the route by which it had entered. Smith made the return trip to San Pedro, a port near San Gabriel, on board the ship *Courier,* commanded by an American, Captain William H. Cunningham.[34]

With fresh horses bought at Los Angeles, supplies and parting gifts from Father Sánchez, the Smith company recrossed the San Bernardino Mountains; but instead of leaving California Smith headed northwest into San Joaquin Valley, where, the trappers had heard, beaver abounded in unexploited streams.

From a base camp on the Stanislaus River, which he called Appelamminy, Smith and his men trapped and explored as far north as the American River. The Mokélumne Indians, thinking them Mexican soldiers, surrounded them, but repented when the American marksmen, at long distance, picked off a few of the boldest. Later the Mokélumnes and other gentile Indians, enemies of the people in the settlements, became friendly with the visitors, and this apparent alliance caused resentment.[35]

In the spring of 1827 Smith made several attempts to cross the great Sierra Nevada, which he called Mount St. Joseph. Losing horses in the snow, his men in danger of death from freezing, he was finally obliged to give up the attempt to take the entire expedition over the range. He then returned to the Stanislaus, there left all his company except Robert Evans and Silas Gobel, and with them set out for the summer rendezvous near Salt Lake to get reinforcements and supplies.[36]

He succeeded in crossing the mountains in eight days, though he lost two of his seven horses and one of his two mules in the passage.

We do not see Jedediah Smith again, so far as the records go, until a month later, when we find him plodding along the southern end of the Deep Creek Mountains, near what is now the Utah-Nevada State line. He had left only one horse and the mule, neither of which had strength enough to carry a man on its back. The condition in which Jedediah and his two men then were is evident from his words in the *Journal* which resumes the narrative at this point.

Marker on Mojave Indian Trail, over which Jedediah Smith traveled when entering California. Photo by Dr. Owen C. Coy, Los Angeles.

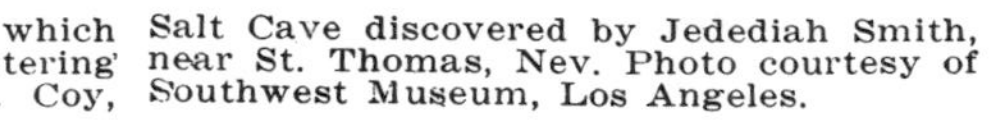

Salt Cave discovered by Jedediah Smith, near St. Thomas, Nev. Photo courtesy of Southwest Museum, Los Angeles.

Marker commemorating Jedediah Smith's visit to San Diego.

Continuation of Journey Across Great Sand Plain

NORTH 25 Miles. My course was nearly parallel with a chain of hills in the west, on the tops of which was some snow and from which ran a creek to the north east.[37] On this creek I encamped. The Country in the vicinity so much resembled that on the south side of the Salt Lake that for a while I was induced to believe that I was near that place. During the day I saw a good many Antelope, but could not kill any. I however, killed 2 hares which, when cooked at night we found much better than horse meat.

June 23[d] N E 35 Miles. Moving on in the morning I kept down the creek on which we had encamped until it was lost in a small Lake. We then filled our horns and continued on our course, passing some brackish as well as some verry salt springs,[38] and leaving on the north of the latter part of the days travel a considerable Salt Plain.[39] Just before night I found water that was drinkable, but continued on in hopes of finding better and was obliged to encamp without any.

June 24th N E 40 Miles. I started verry early in hopes of soon finding water. But ascending a high point of a hill I could discover nothing but sandy plains or dry Rocky hills with the exception of a snowy mountain off to the N E at the distance of 50 or 60 Miles.[40] When I came down I durst not tell my men of the desolate prospect ahead, but framed my story so as to discourage them as little as possible. I told them I saw something black at a distance, near which no doubt we would find water.

While I had been up on the hill one of the horses gave out and had been left a short distance behind. I sent the men back to take the best of his flesh, for our supply was again nearly exhausted, whilst I would push forward in search of water.

I went on a shorter distance and waited until they came up. They were much discouraged with the gloomy prospect, but I said all I could to enliven their hopes and told them in all probability we would soon find water. But the view ahead was almost hopeless.

With our best exertion we pushed forward, walking as we had been for a long time, over the soft sand. That kind of traveling is verry tiresome to men in good health who can eat when and what they choose, and drink as often as they desire, and to us, worn down with hunger and fatigue and burning with thirst increased by the blazing sands, it was almost insurportable.

At about 4 O Clock we were obliged to stop on the side of a sand hill under the shade of a small Cedar. We dug holes in the sand and laid down in them for the purpose of cooling our heated bodies. After resting about an hour we resumed our wearysome journey, and traveled until 10 O Clock at night, when we laid down to take a little repose. previous to this and a short time after sun down, I saw several turtle doves, and as I did not recollect of ever having seen them more than 2 or 3 miles from water I spent more than an hour looking for water. but it was in vain. Our sleep was not repose, for tormented nature made us dream of things we had not and for the want of which it then seemed possible, and even probable, that we might perish in the desert unheard of and unpitied.

In those moments how trifling were all those things that hold such an absolute sway over the busy and the prosperous world. My dreams were not of Gold or ambitious honors, but of my distant, quiet home, of murmuring brooks, of Cooling Cascades. After a short rest we continued our march and traveled all night. The [sound][40a] murmur of falling waters still sounding in our ears and the apprehension that we might never live to hear that sound in reality weighed heavily upon us.

June 25th. [The sun of this day arose on the parched waste and it seemed to us that we were the most unhappy beings on which it poured its floods of light.][40b]

When morning came it saw us in the same unhappy situation, pursuing our journey over the desolate waste, now gleming in the sun and more insuportably tormenting than it had been during the night. [About][40c] at 10 O Clock Robert Evans laid down in the plain under the shade of a small cedar, being able to proceed no further. [We could do no good by remaining to die with him and we were not able to help him along, but we left him with feelings only known to those who have been in the same situation and with the hope that we might get relief and return in time to save his life.][40d]

The Mountain of which I have before spoken was apparently not far off, and we left him and proceeded onward in the hope of finding water in time to return with some in season to save his life. After traveling about [traveling about][40e] three Miles we came to the foot of the Mt and there, to our inexpressible joy, we found water. Goble plunged into it at once, and I could hardly wait to bath my burning forehead before I was pouring it down [in a][40f] regardless of the consequences.

Just before we arrived at the spring I saw two indians traveling in the direction in which Evans was left, and soon after the report of two guns was heard [was heard][40g] in quick succession. This considerably increased our apprehension for his safety, but shortly after a smoke was seen back on the trail and I took a small kettle of water and some meat and going back, found him safe. He had not seen the indians and had discharged his gun to direct me where he lay, and for the same purpose had raised a smoke.

He was indeed far gone, being scarcely able to speak. When I came [within hearing but was not yet in sight he][40h] the first question he asked me was, have you any water? I told him I had plenty and handed him the kettle, which would hold 6 or 7 quarts, in which there was some meat mixed with the water. O says he, why did you bring the meat and putting the kettle to his mouth he did not take it away until he had drank all the water, of which there was at least 4 or 5 quarts, and then asked me why I had not brought more. This, however, revived him so much that he was able to go on to the spring.

I cut the horse meat and spread it out to dry, and determined to remain for the rest of the day that we might repose our wearied and emaciated bodies. I have at different times suffered the extremes of hunger and thirst. Hard as it is to bear for successive days the knawings of hunger, yet it is light in comparison to the agony of burning thirst and, on the other hand, I have observed that a man reduced by hunger is some days in recovering his strength. A man equally reduced by thirst seems renovated almost instantaneously. Hunger can be endured more than twice as long as thirst. To some it may appear surprising that a man who has been for several days without eating has a most incessant desire to drink, and although he can drink but little at a time, yet he wants it much oftener than in ordinary circumstances.

In the course of the day several indians showed themselves on the high points of the hills, but would not come to my camp.

26th June N 10 miles along a valley and encamped at some

Pilot Rock, at the north end of Skull Valley, from the top of which the Salt Lake may be seen

Big Salt Spring, Skull Valley, Utah. One of the disappointments-met by Smith and the "mere skeletons" who accompanied him when he crossed Nevada and Utah. **Photo by Charles Kelly, Salt Lake City.**

Fish Springs, showing Deep Creek Mountains in background, and old pony express station, built long after Jedediah made the first desert crossing. **Photo by Charles Kelly, Salt Lake City.**

brackish water, having passed during the day several salt springs and one Indian lodge. The lodge was occupied by 2 indians, one squaw and 2 children. They were somewhat alarmed, but friendly, and when we made signs to them of being hungry they cheerfully divided with us some antelope meat. They spoke like the Snake Indians and by enquiry I found that they were Pahnakkee's from Lewis's River.[41] They had some pieces of Buffalo Robes and told me that after a few days travel to the North East Buffalo were plenty. Although they knew the Shoshones I could not learn any thing from them in relation to the Salt Lake. In the evening I discovered from a high piece of ground what appeared to be a large body of water.

June 27th North 10 Miles along a valley[42] in which were many salt springs. Coming to the point of the ridge which formed the eastern boundary of the valley I saw an expanse of water Extending far to the North and East. The Salt Lake, a joyful sight, was spread before us.

Is it possible, said the companions of my sufferings, that we are so near the end of our troubles. For myself I durst scarcely believe that it was really the Big Salt Lake that [was before me][42a]I saw. It was indeed a most cheering view, for although we were some distance from the depo, yet we knew we would soon be in a country where we would find game and water, which were to us objects of the greatest importance and those which would contribute more than any others to our comfort and happiness.

Those who may chance to read this at a distance from the [be surprised][42b] scene may perhaps be surprised that the sight of this lake surrounded by a wilderness of More than 2000 [thousan][42c] Miles diameter excited in me those feelings known to the traveler, who, after long and perilous journeying, comes again in view of his home. But so it was with me for I had traveled so much in the vicinity of the Salt Lake that it had become my home of the wilderness.[43]

After coming in view of the lake I traveled East, keeping nearly paralel with the shore of the lake. At about 25 Miles from my last encampment I found a spring of fresh water and encamped. The water during the day had been generally Salt. I saw several antelope, but could not get a shot at them.

28th East 20 Miles, traveling nearly parallel with the shore of the Lake. When I got within a mile of the outlet[44] of the Uta Lake, which comes in from the south East, I found the ground, which is thick covered with flags and Bulrushes, overflowed to a considerable distance from the channel, and before I got to the [channel][44a] current the water had increased to between 2 & 3 feet and the cain grass and Bulrushes were extremely thick.

The channel was deep and as the river was high was of course rapid and about 60 yards wide. As I would have to wade a long distance should I attempt to return before I would find dry land, I determined to make a raft, and for this purpose cut a quantity of Cain Grass, for of this material there was no want. The grass I tied into Bundles, and, attaching them together, soon formed a raft sufficiently strong to bear my things.

In the first place I swam and lead my horse over, the mule following, to the opposite bank, which was also overflowed. I then returned and, attaching a cord to the raft and holding the end in my mouth, I swam before the raft while the two men swam behind. Unfortunately neither of my men were good swimmers, and the current being strong, we were swept down a considerable distance, and it was with great difficulty that I was enabled to reach the shore, as I was verry much strangled.

When I got to the shore I put my things on the mule and horse and endeavored to go out to dry land, but the animals mired [down][44b] and I was obliged to leave my things in the water for the night and wade out to the dry land. We made a fire of sedge, and after eating a little horse flesh, we laid down to rest.

29th 15 Miles North Early in the morning I brought my things out from the water and spread them out to dry. We were verry weak and worn down with suffering and fatigue, but we thought ourselves near the termination of our troubles, for it was not more than four days travel to the place where we expected to find my partners.

At 10 OClock we moved onward and after traveling 15 Miles encamped. Just before encamping I got a shot at a Bear and wounded him badly, but did not kill him. At supper we ate the last of our horse meat and talked a little of the probability of our suffering being soon at an end. I say we talked a little, for men suffering from hunger never talk much, but rather bear their sorrows in moody silence, which is much preferable to fruitless complaints.

30th North 15 Miles I started early and as Deer were tolerably plenty I went on ahead and about 8 O Clock got a shot at a Deer he ran off I followed him and found a good deal of blood and told the men to stop while I should look for him.

I soon found him laying in a thicket. As he appeared nearly dead, I went up to him, took hold of his horns, when he sprang up and ran off. I was vexed at myself for not shooting him again when it was in my power, and my men were quite discouraged. However, I followed on and in a short time found him again. I then made sure of him by cutting his ham strings. It was a fine, fat Buck, and it was not long before we struck up a fire and had some of his meat cooking. We then employed ourselves most pleasantly in eating for about two hours and for the time being forgot that we were not the happiest people in the world, or at least thought but of our feast that was eaten with a relish unknown to a palace.

So much do we make our estimation of happiness by a contrast with our situation that we were as much pleased with our fat venison on the bank of the Salt Lake as we would have been in the possession of all the Luxuries and enjoyments of a civilized life in other circumstances. These things may perhaps appear trifling to most readers, but let any one of them travel over the sand plain as I did and they will consider the killing of a buck a great achievement and certainly a verry useful one. After finishing our repast the meat of the Deer was cut and dried over the fire.

July 1st 25 Miles North along the shore of the Lake. Nothing material occurred.

2nd 20 Miles North East Made our way to the Cache. But Just before arriving there I saw some indians on the opposite side of a creek. It was hardly worth while as I thought, to be any wise careful, so I went directly to them and found as near as I could judge by what I knew of the language to be a band of the Snakes. I learned from them that the Whites, as they term our parties, were all assembled at the little Lake,[45] a distance of about 25 Miles. There was in [the][45a] this camp about 200 Lodges of indians and as the[y] were on their way to the rendevous I encamped with them.

3d I hired a horse and a guide and at three O Clock arrived at the rendezvous. My arrival caused a considerable bustle in camp, for myself and party had been given up as lost. A small Cannon brought up from St. Louis was loaded and fired for a salute.

* * *

My preparations being made I left the Depo on the 13th July 1827 with eighteen men[46] and such supplies as I needed. My object was to relieve my party on the Appelamminy and then proceed further in examination of the country beyond Mt. St. Joseph and along the sea coast. I of course expected to find Beaver, which with us hunters is a primary object, but I was also led on by the love of novelty common to all, which is much increased by the pursuit of its gratification.

112

one of the horses gave out, and had been left a short distance behind. I sent the men back to take the best of his flesh for our supply was again nearly exhausted whilst I would push forward in search of water. I went on a short distance and waited until they came up. They were much discouraged with the gloomy prospect but I said all I could to enliven their hopes and told them in all probability we would soon find water. But the view ahead was almost hopeless. With our best exertion we pushed forward walking as we had been for a long time over the soft sand. That kind of traveling is very tiresome to men in good health who can eat when and what they choose and drink as often as they desire, and to us worn down with hunger and fatigue and burning with thirst increased by the blazing sands it was almost insupportable. At about 4 O Clock we were obliged to stop on the side of a sand hill under the shade of a small Cedar. We dug holes in the sand and laid down in them for the purpose of cooling our heated bodies. After resting about an hour we resumed our wearysome journey, and traveled until 10 O Clock at night, when we laid down to take a little repose. Previous to this and a short time after sun down I saw several turtle doves and as I did not recollect of ever having seen them more than 2 or 3 miles from water I spent more than an hour in looking for water. but it was in vain. Our sleep was not repose for tormented nature made us dream of things we had not and for the want of which it then seemed possible and even probable we might perish in the desert unheard of and unpitied. In those moments how trifling were all those things that hold such an absolute sway over the busy and the prosperous world. My dreams were not of Gold or ambitious honors but of my distant quiet home of murmuring brooks of Cooling Cascades. After a short rest we continued our march and traveled all night. The murmur of falling

Extract from the Smith Journal

I had learned enough of the Sand Plain in my late journey across it to know that it would be impossible for a party with loaded horses and encumbered with baggage to ever cross it. Of the [nine] animals with which I left the Appelamminy but two got through to the Depo, and they were, like ourselves, mere skeletons. I therefore was obliged to take the More circuitous route down the Colorado, which, although much better than that across the [still][46a] Plain, was yet a journey presenting many serious obstacles.

Leaving the head of the Little Lake I moved in a South Eastern direction to Bear River, which, where I struck it, was about yards wide. From Bear River I struck S W to Webers River and up the River Nearly South until I came to where it turned too much S E. I then turned S W across a divide and struck a small stream which ran S W. I followed it down to its mouth in the Uta Lake. Near the Lake I found a large Band of the Uta's encamped.

I[n] coming from the rendezvous to the Uta Lake I was employed about six days. The Uta Lake is . . . .[47]

I traded with the indians for the such things as I wanted, among which was two horses, and leaving the Lake I moved on south to Ashleys River[48] and got on to my old track to Lost River. The Utas had told me of some men that came from this direction last Spring and passed through their country on their way to Taos. They said they were nearly starved to death. On Ashleys River I saw tracks of horses and mules which appeared to have passed in the spring when the ground was soft. These tracks were no doubt made by the party the indians spoke of.

In this time nothing worth relating occurred. At Lost River the indians who were so wild when I passed the year before came to me by dozens. Every little party told me by Signs and words so that I could understand them, of the party of White Men that had passed there the year before, having left a knife and other articles at the encampment when the indians had ran away. I made them some small Presents and moved on to Adams River and down to Pauch Creek and to the indian Lodges on Corn Creek,[49] where I first found corn and pumpkins on my route of last year.

Not an indian was to be seen, neither was there any appearance of their having been there in the course of the summer their little Lodges were burned down. From this place, instead of taking the route I had before traveled through the Mountain by following in the channel of the river, I followed up Corn Creek about 25 Miles and then turning S W I crossed the Mountain without any difficulty, and crossing some low Ridges, struck a Ravine which I followed down to the bed of the dry River which I call Pautch Creek,[50] which I followed down to [the bed][50a] Adam's River about 10 miles below the Mou^tn^.

Since I struck Adams River I had seen but one Indian, and he kept as close to Rock as a Mountain Sheep. I could not account for the indians being so wild. I passed one place where there was a little corn, but it was just in the tassel. Along the Creek I saw several works for making making the sugar or Candy of which I have before spoken.

From the Mouth of Pautch Creek nothing material occurred until my arrival at [Bitter] Creek. I stoped at the Salt Cave[51] and took some salt. On the left hand side of the Creek [and][51a] about three Miles below my camp and ¼ of a mile from the Creek I happened to observe a perpendicular bluff of Salt facing the Creek and like the salt of the cave with the exception, perhaps, that it is not quite as pure. The indians still continued as wild as on Pautch Creek.

From [Bitter] Creek I moved on to the Mouth of Adams River,[52] where I found the old Pautch farmer still on the east side of the Colorado. From this place to the first Amuchaba[53] village my route was the same as when I passed before, with the exception that instead of taking the ravine in which I had so much difficulty I took another further south and passed in to the river without difficulty.

As there had been no indians to carry news of our approach, on our arrival at the village the indians all ran off, but finding an opportunity to talk with one of them, the[y] soon returned and seemed as friendly as when I was there before.

I remained a day to rest my fatigued animals and then moved down to the next settlement. The indians had heard of my approach and met me some distance above their village.

I went to the place where I intended to cross the Colorado and encamped in a situation where I found good grass, with the intention of giving my horses some rest. I exchanged some horses, Bought some Corn and Beans and made a present to the Chiefs. My interpreter, Francisco, who was still there, told me that since I had left there the last summer a party of Spaniards & Americans from the Province of Sonora,[54] by the way of the Gila, had been there. He showed me some things they had got from them. He said the[y] had quarreled and separated, one party going up the Colorado and the other in another direction. This accounted for the tracks of horses and Mules I had seen on Ashley river and for the starved party which the Utas said had passed through their country.[55]

Having traded with the Mojaves, the Smith party began crossing the Colorado, swimming the horses and carrying clothing and goods on a raft. Some of the property had been landed on a bar when the Indians fell upon the almost naked whites, killing ten of them and carrying off as prisoners two Indian women who accompanied them.[56]

Thomas Virgin was injured by a Mojave war club. Eight other survivors either fought their way out of the melée, or fled the unequal contest, or, perhaps, were guarding the landed property at the time of the attack. Smith again takes up the narrative, noting that he and his eight men saw hundreds of savages bearing down upon them.

After weighing all the circumstances of my situation as calmly as possible, I concluded to again try the hospitality of the Californians. I had left with my party on the Appelamminy a quantity of Beaver furs, and if the Governor would permit me to trade, and I could find any person acquainted with the value of furs, I might procure such supplies as would enable me to continue my journey to the North.

But to return from this anticipation, I was yet on the sand bar in sight of My dead companions and not far off were some hundreds of indians who might in all probability close in upon us and with an Arrow or Club terminate all my measures for futurity. Such articles as would sink I threw in to the river and spread the rest out on the sand bar. I told the men what kind of Country we had to pass through and gave them permission to take such things as they chose from the bar.

After making their selection, the rest was scattered over the ground, knowing that whilst the indians were quarreling about the division of the spoils we would be gaining time for our escape. We then moved on in the almost hopeless endeavor to travel over the desert Plain, where there was not the least probability of finding game for our subsistence. Our provision was all lost in the affray, with the exception of about 15 lbs of dried Meat.

We had not gone more than ½ Mile before the indians closed around us, apparently watching the proper moment to fall on us. I thought it most prudent to go in to the bank of the river while we had it in our power, and if the indians allowed us time, select the spot on which we might sell our lives at the dearest rate. We were not molested and on arriving on the bank of the river we took our position in a cluster of small

Cotton Wood trees, which were generally 2 or 3 inches in diameter and standing verry close.

With our knives we lopped down the small trees in such a manner as to clear a place in which to stand, while the fallen poles formed a slight breast work. We then fastened our Butcher knives with cords to the end of light poles so as to form a tolerable lance, and thus poorly prepared we waited the approach of our unmerciful enemies.

On one side the river prevented them from approaching us, but in every other direction the indians were closing in upon us, and the time seemed fast approaching in which we were to come to that contest which must, in spite of courage, conduct and all that man could do, terminate in our destruction.

It was a fearful time. Eight[57] men with but 5 guns were awaiting behind a defence made of brush the charge of four or five hundred indians whose hands were yet stained with the blood of their companions.

Some of the men asked me if I thought we would be able to defend ourselves. I told them I thought we would. But that was not my opinion. I directed that not more than three guns should be fired at a time and those only when the Shot would be certain of killing. Gradually the enemy was drawing near, but kept themselves covered from our fire.

Seeing a few indians who ventured out from their covering within long shot I directed two good marksmen to fire they did so and two indians fell and another was wounded. Uppon this the indians ran off like frightened sheep and we were released from the apprehension of immediate death.

The indians did not press on us again, and just before dark we commenced our Journey and traveled all night and the next morning got to the first spring. As we had no way of carrying water and the weather was verry warm I remained at the spring during the heat of the day and in the evening moved on, traveling all night. In a low plain and in the night when I could not see the distant and detached hills I had no guide by which to travel and therefore lost my way. In the morning I ascended a hill, but could not ascertain on which side the trail Lay.

Observing a high hill nearly in the direction in which I wished to travel I told the men which way to travel in case we did not come back soon, and taking a man with me who was a good walker, I pushed on in search of water and fortunately found some. I sent the man back to the other men and in the mean[time] laid down to take a little sleep, which from my incessant anxiety and fatigue had become quite necessary.

When the party arrived I left them at the little spring and mountaing the highest hill I could see I was enabled to determine that we were about five miles on the right of the trail and nearly opposite a place where I had found water when I passed before. We remained at the spring until nearly night, and then bearing the spring on the trail to the left, I struck directly for the next spring on the old route, traveling and resting by intervals during the night and the following morning until ten O Clock, when we got to the spring.

We there remained during the remainder of the day and the following night, and in the morning early we started, but instead of following the old trail I turned to the left and struck directly for the Salt Plain.[58] My guides had told me of that route when I was there before, but it was considered too stoney for horses.

The day was extremely warm and consequently we suffered much from thirst, my men more than myself, for they had not been accustomed to doing without water as much as I had. We found some relief from chewing slips of the Cabbage Pear, a singular plant which I think I have before described; very juicy although frequently found growing on the most parched and Barren ground.

My men were much discouraged, but I cheered and urged

them forward as much as possible and it seemed a happy providence that lead us to the little spring in the edge of the Salt Plain, for there was nothing to denote its place and the old trail was filled up with the drifting sand. Two of the men had been obliged to stop two or three miles before we got to the spring, and although it was just night two of the men took a small kettle of water and went back, found and brought them up. After dark we proceed[ed] on across the Salt Plain and stopt at the holes I had dug when I passed before and there remained for the rest of the night.

On the following day I moved on to inconstant River. It was still dryer than when I passed the year before. I think it reasonable to suppose that the Salt of the Plain has been formed by a deposit at different times from the overflowing of Inconstant River. The water of the river is sufficiently brackish and the country near the place where it is finally lost in the sand is sufficiently level to justify the conclusion that in some seasons of the year, when the water is most abundant, it spreads over the Plain, and as the dry season approaches the water disappears and leaves a deposit of salt which has in the course of years produced the beautiful encrustation found in the Salt Plain.

About 8 miles up the river I found 2 horses and soon after 2 indian Lodges. I determined at once to secure the horses. And as the indians did not discover me until I had got close to them they had no chance to run off. I found them to be Pauch. With some cloth, knives, Beads &c., which we had brought along, I purchased their horses, some cane grass candy and some demi jons for carrying water.

I then proceeded on, nothing material occuring until I got near the head of Inconstant River there I fell in with a few lodges of the Wan-uma's[59] indians. They had two horses which I purchased, and in continuing my journey, instead of traveling south East around the bend of the stream I struck directly across the Plain Nearly SSW to the Gape of the Mountain.[60]

So soon as I had passed through the Mountain and near the place where I encamped the first night after leaving St. Bernardino[61] on my first journey, I saw numbers of cattle. I immediately determined to kill some cattle and dry the meat to support us in our journey through the Barren country Between Bernardino and the Appelaminy. I therefore had three cows shot and the meat cut and dried.

As the distance to the Mission was considerable I would not go in, but send word to Bernardino of what I had done, but the overseer came out bringing with him such little Luxuries as he had, and as he appeared anxious that I should go in and stay with him a night at the farm house I did so and was verry well treated.

When the overseer came out, among other things he had brought a horse for me to ride in. On the following day he came with me and brought with him some horses which were purchased with things we had brought on our backs from the Amuchabas, which, with those we had before, made each of us a horse.

Two of my men I left at Bernardino, Thomas Virgin, who had been wounded by a club at the Amuchaba affray, and Isaac Galbraith,[62] a free trapper who belonged to my party[63] and preferred remaining, to which I did not object.

The overseer told me that some of the Amuchaba chiefs had been in to the settlement and brought the news of their having defeated a party of Americans, which was no doubt the same of which Francisco spoke. But instead of quarreling among themselves the probability is that they were defeated by the indians, separated in two parties in the affray, and traveled different ways.

In coming from the Amuchabas I had been 9½ days. And remained at the camp 5 days. In the mean time among other things I procured some paper and wrote to father Sanches giving the reasons of my coming to the country and also the reasons that induced me to leave Mr. Virgin. It was my intention that Father Sanchez should advise the Governor of what I had communicated to him.

After having remained [three][63a] five days at the camp near Bernardino I moved off towards my party[64] on the Appelamminy,[65] directing Mr. Virgin, as soon as his health would permit, to come on directly to St. Francisco, for which purpose he was furnished with a good horse. With some small exception I traveled the same route I had passed before and arrived on the Appelaminy and found my party on the 18th of [August][65a] September.

They were becoming somewhat anxious for my return, as it was within two days of the latest time fixed for my return. I was there by the time appointed, but instead of Bringing [the expected Supplies I came to bring them disappointment and defeat][65b] them the expected supplies I brought them intelligence of my misfortunes and the consequent disappointment.

I found them all well. They had passed what hunters call a pleasant Summer, not in the least interrupted by indians. The game consisted of some deer and Elk and Antelopes in abundance. They spoke in high terms of the climate. The air was extremely pleasant from the effect of a gentle North Western Breeze that rose and sank with the rising and setting of the sun.

The indians appeared to be very honest, having at no time manifested a disposition to steal, and entirely friendly. The old Macalumbry Chief (Te-mi), of whom I have before spoken, frequently visited them, bringing them grass seid meal, currents and raspberries &c, and they in return loaded him with meat, which appeared to the indians of this country a most acceptable present.

Among other incidents It may not be amiss to mention that Te Mi brought the stolen horse, as he had promised, a few days after I started for the Depo. He had also brought 7 or 8 of the traps lost in Rock River[66] that had been broken in pieces by the indians, but the men had repaired them.

A party of Spaniards had visited them in the summer, having received intelligence of their being in the country from some indians who had gone in to the Missions. They appeared satisfied with the reasons Mr. Rodgers gave for his being in the country.[67]

I stayed two days with my party, arranged them for trapping, and taking 3 Men with me I started to go in to the Mission of St. Joseph, a distance of about 70 Miles S W. I took some indians with me for guides & on the 3d day arrived at St. Joseph.

I rode up in front of the Mission, dismounted and walked in. I was met by two reverend fathers. One father [Narciso Durán] belonging to the Mission of St. Joseph and the other father [José Viader] of the Mission of Santa Clara. The reverend fathers appeared somewhat confused by my sudden appearance and could not or would not understand me when I endeavored to explain the cause of my being in the country.

They did not appear disposed to hear me, and told me I could go no further and soon showed me the way to the guard house. My horses were [taken] away and for two days I could get no satisfaction whatever.[68] They would neither put me in close confinement nor set me at liberty. No provision whatever was made for my subsistence and I should have suffered much had it not been for the kindness of the old overseer, who invited me at each meal to partake with him. My men likewise ate at the same place.

After 2 days hearing of a man, an American by the name of [William] Welch,[69] I sent for him. He had the kindness to come immediately. I then endeavored to have an interview with the Reverend father [Durán].[70] He condescended to let me know that an officer would soon be up from St. Francisco to enquire what business I had in their country. He asked me if I had anything to eat, thinking I suppose that two or three days was nothing for a heretic to go without eating, as this was the first time he had mentioned the subject, perhaps presuming that I lived on faith instead of food.[71]

I lived in the same way for several days. Finally Lieut. Martinos[72] came up from St. Francisco. After a little conversation with him I found I was to be tried for an intruder and for claiming the country on the Peticutsy.[73] I hardly knew what to say to this charge, but by enquiry I found that an indian had been over on the Peticutsy and returned with his own story of the views of my party.

In the presence of the father, the indian and my self were confronted, Lieut. Martinos sitting as judge. I put a few questions to him by which I ascertained that he had seen me just before my departure for the Depo and had once been with my party during my absence. But no circumstances could be proved against me and Lieut. Martinos instead of punishing me as the father desired Sentenced the indian to a severe flogging, which perhaps he did not deserve.

The father seemed much interested against me for what reason I know not unless perhaps It might be that he was apprehensive of danger to the *true faith,*[74] for which reason he was anxious to stop my fishing around the country (for so he termed my traveling in their country).[75]

I gave the Lieu^{t} to understand my situation and my wants and hinted at my desire to go directly to Monterrey, the present residence of the Governor, for I considered this the most expeditious way to get through with my business. He told me I would be obliged to remain at the Mission until an express could go to and return from Monterey.

Endeavoring to impress him with an Idea of the importance of despatch I urged him to expedite the business as much as possible. He prevailed on the father to furnish me with a room.

After this my meals were sometimes brought to me in my room and sometimes I ate with the overseer at before.

Capt. [John Rogers] Cooper,[76] a Bostonian who had married and resided in Monterrey & Mr. [Thomas B. Park], supercargo of the Brig Harbinger from Boston, came up in company from Monterrey and remaining at St. Joseph 2 days much relieved the anxiety of mind attendent on the uncertainty of my situation. Capt. Cooper in particular seemed willing to afford me any assistance in his power.

I was detained at St. Joseph 12 or 14 days before I received a letter from the Gov. and at the same time a guard to accompany me to Monterey. During this interval I . . . .[77]

On the receipt of the Governors letter I made all haste and started immediately for Monterey. The journey employed us [eight][77a] three days until 11 O Clock at night before we arrived at the Presidio, where I was immediately introduced to the Guard house and closely watched until the next day. In the mean time Capt. Cooper came to see me, bringing some brekfast with him and endeavoring to console me in my unhappy situation.

At 11 O Clock I was informed that the Governor was ready to see me. He met me at the door, shook hands with me and passed a few compliments in Spanish. We then walked through a hall into a Portico, and sit down.

He then commenced talking in Spanish when I immediately told him it would be necessary to have an interpreter. He assented and proposed having Mr. Hartwell,[78] who he said was the only good interpreter in Monterey but who was absent and would not be at home until the evening. He asked me if I would have some breakfast and said he would have a room prepared for me. I thanked him and told him that if he had no objection after breakfast [if had no objection][78a] I would go to Capt. Coopers, to which he assented.

At Capt. Coopers I was received with a hearty and sincere

well come and introduced to Mr. Hartwell. In the Evening I had an interview with the Governor and found him in Monterrey distinguished by the same traits as those that Marked his character when I saw him at San Diego: Nothing further could be concluded on than this. That I should be allowed the privilege of the town, Precidio and harbor. His Excellency appointed an afternoon for an interview at the house of Mr. Hartwell.

I saw at Monterrey Ferguson, the man who ran away from me at St. Gabriel, and Wilson, the man I had discharged at the Chintache lake.[79] At the appointed time I went to Mr. Hartwells and met the Governor. He comenced the business of the interview by observing that what I had stated with regard to my business might be true but he could not believe it for said he, when you came to San Diego you represented the route by which you had come in to California as being a dry barren desert almost impassible, and now you have come by the same route again. It is a verry circuitous route and if, as you say, your only object was to strengthen and supply your Appelaminy party why did you not come directly across to them? And further, when you was defeated and came in so near St. Gabriel why did you not notify me of your arrival?

In answering him I told him it was verry true that when at San Diego I had represented the route by which I had come in as being verry bad, which was a fact, but that on trial I had found the direct route much worse in fact I considered it entirely impassible for a party with loaded horses at that season and perhaps at any other. Of two evils it was natural and politic to choose the least, in doing which I had taken my old route down the colorado. In regard to the notification of which he spoke, I had made a communication to Father Sanches under the impression that it would be forwarded to him immediately.

He did not appear satisfied with my explanation, said it was altogether a misterious business, and that he must have time to consider the subject. From this time I called on him at intervals of a day or two, and after several days he came to the conclusion that I must go to Mexico.

I told him I was ready to go and the sooner he would send me the better. He told me I should go by the first opportunity. A few days after Capt [William G.] Dana[80] of Monterrey, Master of an English Whaler, was ready to sail for Acapulco, the place of debarking for Mexico. I informed the Governor of the opportunity. And he merely said I might go. I soon found that he was not disposed to put himself to any trouble about it. I asked him if he intended that I should go to Mexico as a prisoner and at my own expense. He said most certainly, if I had the privilege of going in a foreign vessel, but if I would wait two or three months a Mexican Vessel would be going to Acapulco when he might perhaps as a favor from the Capt get a passage for me.

I[t] seemed that this man was placed in power to perplex me and those over whom he was called to govern. That a man should seriously talk of making a man take himself at his own expense to prison. That he should talk to me of waiting 2 or 3 months for a passage to Acapulco. I plainly told him that on such conditions I would not go. Capt Cooper, knowing that I had no money, supposed that to be the reason why I refused to go and told me the want of money should not hinder me from going. I thanked him, but I told him I would not see Mexico on the terms proposed by his honor the Governor.

At another interview a few days afterwards Mr. Hartwell, who always appeared quite willing to assist me and whose opinion seemed to pass with the Governor for law, told him he had thought of a way by which I might be let off without his bringing the responsibility on himself. He said "the English Law allowed 4 Masters of vessels in a foreign port in cases of emergency to appoint an agent for the time being who would act as consular agent [for][80a] until the government could be apprised of their proceeding, and perhaps said he the Americans have such a Law."

This seemed to please the govenor and he said he would see what could be done. No sooner was the conference ended than I told Cap[t] Cooper of what had passed and also the Masters of the several vessels in port. They were not perfectly satisfied of the legality of the proposition, but thought the urgency of the case would justify the proceeding.

Cap[t] Cooper was appointed agent by the different Masters in port [among which was][80b] Having lost my journal of that date I am unable to give the names of but two of them, Capt. [Joseph] Steele of the Brig Harbinger and Capt. [Allen] Tilton of the Ship Omega.

Previous to this the Gen[l] [81] had requested me to write to my party to come in. I told him they were nearer St. Francisco than that place and he remarked that they might go in there. I therefore wrote to Mr. Rodgers that it was the Governor['s] request that they should come in, and at the same time hinted at the treatment I had received. This I knew was sufficient for Mr. Rodgers, who from what had [frequently][81a] passed between us would go in to [San][81b] Bodega.

I carried the letter to the Gen[l] unsealed. He had it translated and took a day or two to consider its contents, then sent for me and said he was afraid to [that letter][81c] send such a letter for I had not ordered Mr. Rodgers positively to come in and that I had discouraged him from coming in from the manner in which I had spoken of [my][81d] the usage I had received at the same time he observed he would be verry sorry [to][81e] that his soldiers should have any difficulty with my party.

I told him I thought what I had written verry reasonable but that if he would give me a copy I would write again. He said he could not do that. After getting the promise of the Govenor that they should not be imprisoned and should be furnished with provision I wrote to Mr. Rodgers directing him to come in to San Francisco. The soldiers who carried the letter went by way of St. Joseph and one of my men accompanied them.

Notwithstanding what the Gen[l] had said about his soldiers and the smallness of my party I think he did not wish to have my party try their rifles on his soldiers, for there was some terrible stories in circulation about the shooting of my men. It was said they were sure of their mark at any distance.

In the mean time news came in from the South that another party of Americans were near Too Larra Lake. I told him I was well convin[c]ed there were no Americans there, but as it was his request I would write to them.[82]

After Cap[t] Cooper was appointed agent the Gen[l] wished him not only to become responsible for my good conduct until I left California but also to insure that I should [not] return again to the country on any pretense whatever. I would not agree to such a restriction and after a short contest the Gen[l] consented to drop it.

I received a letter from Mr. Rogers informing me of his arrival at St. Francisco. I got permission to write to M[r]. Virgin and the Gen[l] agreed to forward it to him.

November 7th 1827 I called on the Governor in company with Capt. Cooper who gave the Gen[l] a written certificate stating the reasons which he thought had brought me to the country, which nearly accorded with what I had stated to the Governor. It also stated what I was in want of, in the mean time offering to become responsible for my conduct.

The Gen[l] said on those conditions I could take my choice of three things either to wait until he could receive orders from Mexico. Or I might go there as an opportunity would offer in 8 or 12 days or I might go away with what men I had in the same direction by which I had come in. He insisted that I should travel the same route by which I arrived and in preventing me from hiring more men he calculated I would be afraid to travel with the number of men I had and consequently he would retain me in the country until he could receive orders from Mexico. But I told his excellency I would go if I had but 2 men. The Genl said he would make a memorandum of what Capt. Cooper must become responsible for and would then call on us.

The Ship Omega, Capt Tilton, departed and in the evening I received a letter from Mr. Rodgers stating that by the help of Capt Richardson of St. Francisco he had got permission to remove to a spring not far distant from the Precidio. This letter came by the hand of Charles Swift and further informed me that the party were well supplied with Beef Corn Beans &c.

9th November & 10th The Genl was sick so that no business could be done. A ship being in the harbor from Boston, [John] Bradshaw[83] Capt, and [Rufus] Perkins Supercargo, I made a contract with the Capt selling him all my furs at $2.50 per lb. On the 10th the Genl gave Capt. Cooper a Copy or pattern for a Bond which he with some little alterations greed to sign. In the first place the Genl wished to obligate me to remain North of the 42nd parallel of Latitude. But he finally satisfied himself to [bind][83a] make Capt Cooper [responsible][83b] guarantee that I should not hunt on the sea coast south of the 42nd parallel of latitude but within Land wherever my Government might permit. Before the Bond was signed the Genl again proposed sending me to Mexico but I told him as I had sold my furs and made arrangements for traveling homewards I could not well do otherwise, and he with some little hesitation assented. We were to have three [origine][83c] copies of the original bond, one to send to Mexico one to be left with the Genl [and an][83d] one for Capt Cooper and one for myself.[84]

12th November The Bonds were all made and signed.[85] The Genl then requested a list of such things as I wished to purchase which I gave him. He objected to none of the articles except Horses and Mules but after some difficulty he allowed me permission to purchase 100 Mules & 50 horses.

15th having got my passport[86] I went on board of the Franklin, Capt. Bradshaws Vessel, and at 2 O Clock we sailed for St. Francisco. I was soon sea sick and a gale of wind that came on at 5 O Clock Made me much worse. At tea time a sudden lurch of the vessel threw the whole apparatus of the table into one of the lower births and Capt Bradshaw received the whole of the Tea but no other damage was done save the breaking of some dishes.

16th At night we were off the Bay of St. Francisco but the wind being contrary we stood off and on until the next day.

17th At 12 O Clock the wind being fair we entered the harbor there were 7 Sail in at the time. I called on Don Lewis[87] the Commandant he seemed satisfied with my passports. I found my men all well but they had not[88] been well supplied with provisions—Mr Viermont[89] (a German) trading under the Mexican flag had been very Kind to them.

18th I had my furs taken on board the Franklin amounting to 1568 lbs of Beaver & 10 Otter Skins. Rainy weather. I was engaged until the

22nd in preparing my good[s] & going back and forth from the ship and party to the Precidio. I made arrangement to have some of my things sent in the Launch to St. Joseph and some put on shore at that place. In company with Capt Bradshaw I received an invitation to dine on Board the Sloop of War Blossom.[90] But was detained so long by my business with Don Lewis that I could not attend.

23^d My things were all ready to go on shore and I received an invitation to dine. The Company consisted of Capt. John Bradshaw of the Ship Franklin from Boston. Capt. Reuben Cresy of the Sophia from New Bedford, Capt. [John] Fo[r]-ster of the Brig Tullum[91] from Mexico, Capt Moses Harris of the Weymouth from Nantucket, Capt [Obed Swain] of the Enterprize from Nantucket, Capt [Benjamin A. Coleman] of the Eagle from Nantucket, and Mr. [Henry D.] Fitch, supercargo of the Brig Tullum. After dinner the wind arose so that I could not go on shore, so we remained and supped and contrary to my wish sat up untill 3 O Clock to drink wine, after which we took a little sleep.

24th Early in the morning I went on shore and as I could not get my work done at St. Francisco I got permission from Don Lewis to remove to St. Joseph as that place would not be out of my way.

26th Since the party had been at St. Francisco the horses had been nearly starved but I got them up and moved off in the direction of St. Jose, about 20 miles, and put them up at the farm of Don Lewis, where I was politely treated at his expense.

27th I went ahead of the company as far as the Pueblo with Mr. Garnier,[92] a man I had engaged to go to Monterrey with two men and two horse loads of Merchandize to pay for some horses I had purchased at that place. At the Pueblo I made application at the house of Mr. Welch for the payment of [a] draft of 200 Dollars on him from Mr. Perkins. But I was disappointed he was not at home and had [sent][92a] left orders for me to send to the Mission of Santa Cruz. I hired a man, sent him there but got no money.

28th I continued on my way to St. Jose, Mr. Garnier being with me for an interpreter. I mad[e] an arrangement with the Priest by which I was to have the use of the Smith Shop for one week for the purpose of repairing my guns and a room for myself and two small rooms for my men. The party arrived at 5 O Clock having left some of their horses behind which had given out.

29th I sent two men down to the Pueblo to bring up any horses which Mr. Garnier might purchase there.

30th My two men returned with 3 horses and 3 Mules sent by Mr. Garnier. I spoke to Capt Bradshaw who had come up with some things for the Priest and for myself of the disappointment I had received in relation to the money from Welch. He told me if I would go down to the Ship he would pay it.

Dec. 1st I started to go down on the north side of the bay in Company with Capt Bradshaw having procured a guide from the Father and a horse from the Overseer. At 4 O Clock I arrived at the Farm of St. Pablo[93] and procuring a Boat and men at 10 O Clock I embarked, leaving Capt Bradshaw. At this time it was flood tide and the wind being being verry high shortly after starting we was obliged to come to under the shelter of a small island for a short time when were enabled to proceed on again and at about 3 O Clock arrived on board the ship.

I procured some such things as I wanted and remained all day waiting for the arrival of Capt. Bradshaw who did not arrive untill 3 O Clock at night. In the morning I got my money and was ready to sta[r]t back, but the wind blew so hard I was obliged to remain until the following morning when I started early and immediately on my arrival at the farm of St. Pablo I mounted my horse and rode up to the Mission. In my absence Mr. Rodgers had driven up the horses to brand and found that five of them were lost.

December 5th My men were Busily engaged in Baleing up my goods for the journey.

6th Dec My men engaged as before. on having my horses driven up I found 6 of them were missing.

Isaac Galbraith, the man who had stayed at St. Bernardino with Mr. Virgin came but brought no news from him. I received Letters from Capt. Bradshaw, Capt. Cooper, Mr. Garnier and Mr. Perkins. But Mr. Perkins said not a word about the money I was to receive from Mr. Welch. I also recd one from father Louis[94] of Santa Cruz stating that on the receipt of my letter he had sent me $150, being all the Money he had at the Mission at the time, but that just after the Messenger started on his return to me he had received a letter from Mr. Welch which caused him to send after the Courier and detain the Money.

7th Palmer & Reed were at work on the Guns and the rest of the party employed as the day before. My horses were all found but 3. I wrote to Don Lewis and to Capt. Bradshaw.

8th Was a Saints day and of course little could be done.[95]

9th Was Sunday. I attended Mass. The Music consisted of 12 or 15 violins 5 Basc vials and one flute. The father spoke in Latin and in Spanish and a part of his discourse was then translated for the indians into their own tongue.[96] Not only on Sunday but every day of the week the indians are called to prayers at an early hour in the church.

10th 11th & 12th My men were employed in the several kinds of preparation for the journey. As yet no word received from Mr. Garnier.

13th I had some of my men engaged in drying Meat and others at work in the Shop. In the course of the day Mr. Virgin arrived. He had been imprisoned for some time and frequently without anything to eat and strictly forbidden to speak to any one, [8 days before][96a] and abused in almost every way. On the 5th the Genl on his way to St. Barbara saw him, released him and instructed the fathers to forward him on to St. Joseph. He was much rejoiced to see us and I am sure I was quite glad to see the old man again.

I am informed by good authority that my young indian of whom I spoke the first time I was in California and who was in prison when I went away was tried for his life charged with having piloted me into the country and sentenced to be shot. But father Sanches, influenced by his own good feelings and his promise to me, wrote to Mexico and procured his pardon. And further I have good reason to believe that the Amuchabas were instructed to kill all Americans coming in in that direction; but let that be as it may the fact that they punished an indian for being friendly to me would readily convey the idea that they would reward them if they were enemies.[97]

15th Some of my men were engaged in cutting meat to dry, others at Blacksmithing and some Loo[k]ing for lost horses. I had spoken to Capt [W. A.] Richardson[98] of St. Francisco when on his way to Monterrey to speak to Mr. [Paul Shélikof] the Russian agent at Bodega who was at that time at Monterrey of the probability of my passing by Bodega on my way north and that I might want to repair my things or procure some supplies. I received a letter from Mr. [Shélikof] informing me that he should soon pass to the North himself and would give instructions to the agent at Ross the Precidio of the Russian Settlement to provide me with whatever I stood in need of.

16th Sunday. I again attended Mass.

17th I had several of my men out hunting horses and one was found. No news as yet from Garnier. My preparations for a start are nearly complete.

18th Arthur Black came from Garnier with the intelligence that he [was in want of money and][98a] had purchased 180 horses and mules and was in want of money. He was at Castros farm[99] 20 Leagues from St. Joseph and from what Black told me I considered it absolutely necessary that I should be on the spot. I soon fixed for starting with 4 men and as my horses were inconvenient I bought some for the trip it rained and snowed considerably but that did not stop me and at 12 O Clock at night I arrived at Castros farm. Since Black had left there the horses had broken out of the pen and several I knew not how many were lost.

19th rainy but I had men out hunting for the lost horses and some of them were found and put in with the band at night carefully guarded for I find that notwithstanding the small value of horses they are frequently stolen.

20th I started with 2 Spaniards and 5 of my own men to drive the band of horses in to St. Jose in the mean time Mr. Garnier and one Spaniard remained behind to look for lost horses. I went on with the Band of horses about half way to St. Jose when I stopped them and pushed on ahead. On my way I hired a young man an Englishman who had been in [the] country about 2 years and was an excellent horseman, his name was Richard Leland.[100]

21st I had my horses which were near St. Jose driven in and started to meet the band, in the mean time intending to [buy] some things of Mr. [John] Burton[101] in the Pueblo and some Blankets of Father Joseph at Santa Clara. I met the Band, did some of my business and returned to St. Jose. In the evening I went again to the Pueblo and as the rain increased I staid all night at Capt Burtons.

22nd I returned to St. Jose and got permission from the father to remove my Company to a sheep farm belonging to the Mission called St. Lorenzo,[102] where there was a plenty of grass and a pen in which I could shut up my horses & Mules.

23^d^ As it was rainy I did not move.

24th I started the Party off to the farm & taking Laplant with me I went as far as Lieut. Martinos on my way to St. Francisco. It was late when we arrived at that place and we remained all night.

25 After taking some of the Lieut's Honey we proceeded on and arriving at St. Francisco I found that nothing could be done on that day, one of Don Lewis's Children having died the night before. All my preparations being completed for moving of[f] to the North I was anxious to be off as soon as possible.

26th I had 2 interviews with Don Lewis. It was the instruction of the Genl to Don Lewis that I should cross the Buenaventura River[103] near its entrance into the Bay of St. Francisco. Don Lewis was further ordered to send 10 Soldiers to see me safe out of their territories. The time which the Genl had given me to remain was nearly expired but I found it entirely impossible to procure a Launch to take me across the river without which it was impassible. The only Launch in the neighborhood belonging to Capt Richardson was unfit for service.

In this situation I made no doubt that Don Lewis would consent that I should go up the River until I could find a place where I could swim my horses and carry my goods over on a raft which could not be done at the mouth. But he would hear nothing of this proposition but insisted that I should cross at the particular place directed by the Genl. I then told him to furnish the boat and I was ready to cross. This he could not do but said I must wait untill the Genl could be advised of the situation of things and give further instructions. I apparently acquiesced but left him with a determination fixed to take my own course without waiting for their tardy Movements which the situation of my finances would not permit.

By riding until 12 at night I arrived at the Pueblo where I found Mr. Garnier. He had got but one of the lost horses. I settled with him and found that in the time he had been employed in Purchasing for me he had lost 19 horses and Mules. I sold them to him for 25 Dollars.

27 I went home to my company and as I am about ready to leave the Settlements of California It may perhaps be appropriate to insert in this place the remarks I have made on the country in the vicinity of the Bay of St. Francisco. The Bay of St. Francisco having been well described by Vancouver I will merely observe that it is universally considered the most safe harbor on the Western Coast of America. It is spacious and has suficient debth of water for the largest vessels. The entrance is safe and about [2] miles wide and the surface of the Bay is protected from the violence of Western winds by a chain of hills that run through the two projecting points of land that form the neck of the Bay. The form of the Bay is somewhat triangular, one of the arms running to the S E and one to the NorthEast. That to the South East is the longest, ex-

tending for some considerable distance, perhaps 20 miles, into the rich valley of the Missions of St. Jose and Santa Clara. The N Eastern and shorter arm is that which Receives the Buenaventura River.

The Precidio of St. Francisco is on the narrow Point of land that forms the S Western Boundary of the Bay and about ½ miles from its entrance & immediately on the shore. There is but little good land in the vicinity of St. Francisco a chain of hills that run parallel with the coast of the Ocean and shore of the Bay come in Close to the Precidio leaving but a narrow strip of fertile soil along the Bay. The Buildings of the Precidio are according to the common custom of the Country built of unbur[n]t brick. They are like Barracks and built in a square and were once capable of accomodating 20 families and 100 Soldiers but are now much decayed.

The entrance of the harbor is defended by a fort placed on the point about ½ mile from the Precidio in a situation admirably adapted for the purpose for which it was intended. It mounts 15 or 20 pieces of cannon but I am told is somewhat decayed. About 3 miles from the Precidio is the Mission of St. Francisco which is not as rich as some others in the vicinity.

Leaving St. Francisco [the Chain][103a] and proceeding S. E. along the Shore of the Bay the Chain of hills on the S West on which are fine forest of cedar gradually retire and leave a fine country pleasingly varied by prairae and woodland. In this delightful country is the Mission of Santa Clara. Not far from the Southern extremity of the Bay and nearly opposite is the Mission of St. Jose.

There is a considerable Stream[104] that enters the South Eastern Extremity of the Bay after winding through the fertile valley Laying Between two Chains of hills one of which ranges nearly North and South and terminates near the Mouth of the Buenaventura and the other of which I have before spoken ranging in a South Western direction from St. Francisco.

[Near the][104a] From the S. E. extremity of the bay extends a considerable Salt Marsh from which great quantities of salt are annually collected and the quantity might perhaps be much in creased. It belongs to the Mission of St. Jose.

About one Mile from Santa Clara is the Pueblo which consists of about 100 houses built of the common material. Unburnt Brick. But few of these are any wise respectable in appearance—the remainder are merely huts. Along the east side of the Bay a fine country extends to the Mouth of the Buenaventura River including several fine farms and among the rest that of Santa Ana near the Mouth of the River. On the North Side of the Bay a fine country is spread nearly to the Ocean where there is a chain of Rocky hills nearly on the coast.

The Mission of St. Raphael is about Two miles from the Bay and nearly opposite the Mouth of the river. It is a fine country which extends indefinitely from the Chain of hills near the coast along the shore of the Bay and up the Buenaventura River. The farm of Santa Anna extends along the shore of the Bay about three miles and back into the country about the same distance. There is verry little land in cultivation and the amount of stock is small for that country, but the soil is excellent and the situation combining many advantages is at the same time Most delightful and pleasant.

The Best farming establishments in California are the Missions. At Each of these there are fine orchards and gardens. Individuals of this country are generally too indolent to make good farms they rather prefer the less laborious task of raising horses and cattle and in this business they are so forcibly assisted by the peculiar advantages of the country that the herds of Cattle have accumulated untill they are nearly as numerous as the Buffalo of the plains of the Missouri and the horses are in many places so plenty as to lead to that barbarous custom of which I have before spoken of shutting them up by hundreds in Parks to Starve. Because they eat the grass from the tame Bands.

I coul[d] not say much of the enterprize of these people. I have heared them speak of the chalk mountain of the east which is no other than Mt. St. Joseph they having never been sufficiently near to [distinguish][104b] determine that what they thought to be [Snow][104c] Chalk was in reality nothing more nor less than Snow.

They have frequently spoken of a party of discovery that went out three or four years since under the direction of Don Lewis. They say they went a long distance up the principal Branch of the Buenaventura or Piscadore and that there was a man in company well acquainted with the Columbia who told them that this river was a Branch or rather Bayou of the Columbia & that [left it something like 600 Miles from the Bay of St. Francisco][104d] they were in a short distance of that River.

This proves to me their enterprize for I well know that it is not over 250 m from San Francisco to the place where the river leaves the Mount St. Joseph. Among other things they described a singular hill in the plain which was near the place where they turned back. By the help of this I was enabled to fix the limits of this Memorable voyage of Discovery that left it doubtful whether the summits of M't. St. Joseph were chalk or snow.[105]

28th & 29th I had my men engaged in breaking mules for the loads. I wrote to the Gen[l] and to Don Lewis informing them of what I intended to do and in the mean time settled off with the father under pretence of moving to better grass.[106]

30th Rainy but I started and made 8 Miles N E and at night had my horses guarded by men on horseback. On the 31st it was Rainy and consequently Muddy but I moved 10 Miles North East.

January 1st 1828 20 Miles East and encamped on Buenaventura River[107] which sometimes is called by the Spaniards the Piscadore. One of my best mules which was tied broke loose and ran away and was lost.

2nd 4 Miles S E and encamped again on the bank of the river which I suppose to be the Peticutsy.[108] I made rafts for crossing of poles and flags.

3d I made a pen on the bank of the river and driving my horses in by small bands into the pen and from the pen into the river I crossed them over without the loss of any, contrary to my expectations.

4th Some of my men were out hunting for Beaver sign and as the water was high the weather rainy and the banks of the river Low I thought it advisable to build some Skin Canoes which would assist us in trapping and in crossing streams in our course.

Having been so long absent from the business of trapping and so much perplexed and harrassed by the folly of men in power I returned again to the woods, the river, the prairae, the Camp & the Game with a feeling somewhat like that of a prisoner escaped from his dungeon and his chains.

7th Some of my men are engaged in hunting Elk for the sake of the skins to make canoes and a few were trapping but I could not do much at trapping for I had but 47 traps. 9 men were attending the traps and the rest of the party not hunting were taking care of the horses and camp keeping. In the course of the day I moved 8 miles down the river and encamped on a creek not far from the river. Nearly all the Lowlands along the river were inundated. At the camp from which I moved I left 4 men to finish two canoes which were nearly done and start down the river trapping. They were to join me again in a week.

10th It had been raining almost every day since I came to the river, and finding that verry little could be done by horses when the rivers were so high I concluded to make another Skin Boat.

11th & 12th Good Weather. I had at that time taken 45 Beaver. I had the skins dried and started 2 men with another canoe. Some of the Appelamminy indians visited me; they were as usual friendly. My horses had eaten the grass so much at [*sic*] make it necessary for me to move my camp.

13th N Westerly 4 Miles and encamp on a creek which was dry when I was in the country the last summer but now had plenty of water.[109] In the vicinity and at this season of the year it was impossible to go [to] the river with a horse for several miles above and below my camp in consequence of the low flaggy ground which was covered with water. Some of the Ponds have Beaver along their flaggy banks and three of my men who trap by land succeed[ed] in taking some of them. My number of Beaver had increased to 61. The weather good.

14th I sent Mr. Rodgers with 2 men to hunt for Elk with instructions to remain out one or two nights as circumstances might require. My 4 Men who went with the first 2 Canoes came in at night bringing with them 33 Beaver. My Men in camp were engaged, some in stretching Beaver skins, one in saddle making and some in breaking Mules. Weather still pleasant.

16th Mr. Rodgers and the 2 men returned having killed 4 Elk. With them came 2 indians. One of them was the principal chief of the Machyma Band of indians that reside on the head of the Mackalumbry River. I was at his village the last spring. I made them some presents and told them I was going to Bodega and they agreed to go with me.

17th Reed & Pompare[110] the 2 Trappers that went in a Canoe by themselves returned having caught 22 Beaver. The two chiefs left camp and said they would join me on Rock River in five days. 2 of the 4 Canoe trappers came in they had put ashore 6 Miles below the river at that place had overflowed its banks so that there was no chance for trapping. As there was now opportunity for trapping by water I directed the men to take horses, go on by land and join me at Rock River.[111]

18th I started for Rock River. Reed & Pompare started in the canoe to Join me in 8 days [hence][111a] on Rock River. In 4 miles I came to the Mackalumbry River[112] and in attempting to cross it I got my horses nearly all mired and was obliged to relinquish the idea of crossing at that place. After some difficulty I got my horses out of the mire and encamped.

19th N E 10 Miles I moved on up the Machallumbry endeavoring to find a suitable [place] to cross. The traveling was verry miry and I continued on ten miles before I found a place that would answer the purpose. I then encamped. In the morning I saw a Grizzly Bear and shot at him but did not kill him.

20th I[t] took me all day to get my goods and property over the river. I fell trees across the water and carried the goods over. We then undertook to cross the horses by swimming them, but could not get them into the water until I had made a pen on the Bank for that purpose. Before they were all over the place where they came out on the opposite bank became so miry that I had to Bridge it. One of the men shot several times at a Bear but did not kill it.

21st 10 Miles North to Rock River Although the ground was rolling the horses sank at every step nearly to the nees. In the morning the 4 Canoe trappers joined me. They had followed up the Buenaventura to the mouth of Rock River and up it several miles but not finding me they struck across to the Mackalumbry and fell in with me just as I was starting. They had seen a great many indians but found them friendly. In the evening a good many indians were seen near our camp, but they all ran off.

22d I sent some men out early to hunt for deer as there was sign in the vicinity and we were destitute of provision. Several indians came to camp and I gave them some tobacco. They brought with them some fine salmon some of which would weigh 15 or 20 lbs. I bought three of them and one of the men killed a deer. In the mean time some of the men were up and some down the river searching for Beaver sign. They found but little and set but few traps. One of the Chiefs that had promised to meet me at this place and accompany me to Bodega came and told me that the other was sick which would prevent their going with me. It rained most of the day and all night.

23 It had been my intention to ford the river and go down on the other side. But the late rains had raised the river so much as to render this impracticable. I therefore remained in camp and although the rain continued the river commenced falling. Many indians visited the camp among the rest some of them who were so hostile to me the last spring. They were quite friendly and I gave them some Tobacco and Sugar. An indian which came the day before pretended he would go with me 2 or 3 days and as he was quite naked I lent him a Blanket. At night he ran off taking with him the borrowed Blanket. There was in camp several indians of the same band and among the rest one who called himself a Chief. I told him of the theft of the Blanket by signs which he understood. He remained at camp and after sending out twice brought in the Blanket. These indians are nearly naked and have less modesty than any I have ever been with.

24th I went down the River 2 Miles, found a ford and crossed over and went 7 Miles further down course nearly W N W. found some Beaver sign set traps and encamped.

25th As I had some Beaver skins which need drying I did not move camp. But took 4 men with me and went down the river as this was the day I was to meet Reed & Pompare. I went about 12 miles and supposed I was near the Main River. But could get no further on account of the miry Bottoms. I therefore left two notes for them. One on the bank of the River and one on the trail and returned in doing which I saw several indians who ran off.

26th I crossed over the river at an indian village of 50 lodges they made a flag canoe to assist me in crossing. I then got an indian who could speak a few words of Spanish and taking him and Laplant with me I went down Rock River 3 Miles further than I had gone the day before and in view of the flags of the Buenaventura. But Rock River turning N W I supposed from the instructions I had given Reed & Pompare that they had passed above its mouth and were then on the main river. As the indian said I could not go up that way on account of the Mire I concluded to return again to the village and send some indians after the 2 men. So I returned and remained all night at the village where I was treated with great Kindness.[113] In the morning I hired six Indians to go [in] search of the lost men & sent them off. I invited the Chief to visit me & returned to Camp.

27th I returned to my camp and soon after the chief with 20 or 30 indians came to see me and I gave them some small presents. These indians called themselves Machalunbrys. I saw among them a good many Spanish Blankets & Shirts. They did not manifest a disposition to steal which as I have before remarked is no small merit for an indian. The[y] have a kind of preparation resembling the Persimmon bread made by the indians on the Lower Mississippi but their principal living seems to be Acorn Mush. Their Lodges are stationary and made of grass and Mats. When the indians came to my camp they Brought a large Basket filled with Mush and asked me for some meat which they ate with their Mush.

28th The Indians I had sent for my 2 men returned without finding them. They said they had seen their tracks where they had been traveling towards Rock River. My horses caused me a considerable trouble and one of my horse guard in looking for horses not far from camp lost himself, Mired his horse and left him. I was absbent and Mr. Rodgers sent some men to look for the horse but the[y] returned without finding him.

29th N W 9 Miles down Rock River. Some horses were missing but I did not wait for them. As soon as it was light I took 4 Men and started on the trail made by the Man who lost [his] horse when he went out. Traveling every point of the compass through Mud and for 7 or 8 Miles I found the horse not more than 3 Miles from camp. By ten O Clock I was back again to camp. 4 Men sent for lost horses returned they had found them but could not get them to camp.

30th I sent three men back after the lost horses. It rain[ed] considerably, but I had my packs put upon the horses & moved down the river & encamped where 4 of my men sent down to trap had encamped last night. In the evening the men sent for the horses returned with part of them, some were still left.

31st I sent two men over to the river to look for Reed and Pompare and I with one man went again down on the North side of Rock River for the same purpose. 2 men were sent for horses. At night all came in. The men from the Buenaventura saw no sign of the lost men. I went with difficulty as far as the Mouth of a stream coming in from the North which I supposed to be indian River. The country miry and the River high. No sign of the lost men. The [men] who went up the River found no horse.

Feb[y] 1st 1828 I took 3 Men and went in search of the horse but did not find him. Two Men were sent down on the oposite side of the river in search of Reed and Pompare with instructions to go if possible to the mouth of Rock River. Four indians visited the camp and one who had been with us several days ran away taking with him a Blanket. My trappers took a few Beaver each day.

2[d] 3[d] & 4th remained at the same camp. The indians brought me word that they had seen my two men some distance above on the Buenaventura. I immediately started two men off in Company with some indians to see if they could be found.

5th I crossed over the River carrying my goods in two skin canoes I had lately made and swimming the horses and then went down the River 4 Miles and encamped.

6th I had another Skin Canoe made. In the course of the [day] 50 or 60 indians visited the Camp. They were from Elk Creek and Indian River and seemed to want nothing but meat which our luck in hunting enabled us to give them as Mr. Rodgers had killed 4 Elk and myself 5. The two men sent to look for Reed & Pompare returned. The indians had taken them in towards the Buenaventura but as it was constant wading they were obliged to return. I sent 6 Men in 3 Skin Canoes down Rock River for the purpose of trapping.

7th A good many indians stayed in camp the last night. The trappers sent down the river returned in consequence of hearing a good deal of shooting and observing some movements of the indians which the[y] did not like. I went down with them again and encamped with them.

8th I went down the River a few miles from the camp of the trappers to see if I could cross the River when ready to move on North, I found a suitable place and returned.

9th I went hunting for Elk but did not kill any. When I returned I found about 60 indians at camp. Ten indians who I had sent two days before in search of Reed and Pompare returned without hearing any thing from them. I then gave up all hopes of seeing them again. On examination I found they had taken nearly all their things with them, from which circumstance I Judged their absence was voluntary and that when they went away they had no intention to return.[114]

10th I moved down the river to the place where I intended to cross. My men thought I could go no further but the indians said I could go on. There were at my camp several of the neighboring indians and I made them some small presents.

11th I went with one man across the river to examine the country on the opposite side. Considering it passible I sent a note back to Mr. Rodgers [to have][114a] directing him to cross over and I continued out into the prairae pursuing Elk but did not kill any.

12th I moved N W 10 Miles across a flat muddy country sometimes in mud and water 2 or 3 feet deep and encamped on the East bank of the Buenaventura River. At that place the River was about 300 yards wide, a gentle current and apparently deep, the water somewhat muddy. The Banks generally low are timbered with Ash Cottonwood Elk Sycamore Willow and where the ground is sufficiently dry some Oak. The Timber on each side of the river is narrow and the boundary between the upland and river bottom is not marked by steep Bluffs, the inclination of the face of the country being on both sides gently sloping towards the river. In the bottoms are Lakes and flags which frequently extend 2 miles from the river. The soil of the country is generally good being frequently a rich chocolate colored loam.

The winter in this valley is the best season for grass and at the time of which I am now speaking the whole face of the country is a most beautiful green, resembling a flourishing wheat field. I have frequently had occasion to speak of the miry muddy traveling which so much obstructed my progress. This is much owing to the superabundance of rain during the winter which in addition to the vast volumes of water poured out of Mt. Joseph fills the streams to overflowing and completely saturates the light rich soil of the valley. I had by means of the residence of my party in the valley during the summer and by my own observation an opportunity to determine that the country was generally sufficiently dry for cultivation except during the rainy season. On [arriving at][114b] the river I found verry little Beaver but plenty of Otter sign. On my first arrival at the river I was under the impression that it was Wild River[115] But soon after I was undeceived.

13th N 12 Miles Some of the [Indians] which came with me as guides from Rock River continued on with me. The Country like the last described. Encamped just below a creek on which was considerable beaver sign and the traps were set.

In the course of the day I passed an Indian village of 20 or 30 lodges made of flag mats and straw. The men were there themselves but had taken the precaution to send off their wives, children and goods. Our arrival caused a good deal of uneasiness but when my guide, the old chief, came up and spoke to them all was well again.

14th Only 6 Beaver were taken the last night. Some Raccoon and 2 Deer were killed but the deer were poor which is always the case in this valley although the grass is good.

15th Was rainy. I had my things all carried over the slou on a log and my horses swam over. The old chief, my guide, left me making signs that he would return. 2 Beaver taken.

16th A circuitous route of 10 Miles following a bend of the river to the west and then North, making a direct line of six Miles. Country dry along the river and flag Ponds back a short distance. Opposite camp the river was much smaller and and as I could see about 2 miles below camp an arm of timber extending off North West I concluded that the main river ran in that direction. South East from camp the timber extended farther from the river and the land became dryer. In the course of the day I passed several indian villages built of flag Mats and straw. But the inhabitants had taken the alarm and fled. Much more Beaver sign than Below. The day was tolerably pleasant.

17th Remained at camp. After 9 O Clock it was Rainy. 14 Beaver taken.

18th Remain[ed] at same Camp. The rain that commenced the day before continued without intermission for 24 hours. This may be well termed the rainy season for we scarcely had more than one pleasant day at a time, 12 Beaver taken.

19th The rise of water interfered so much with trapping that but 6 Beaver were taken. One of my men, E. Lazarus, had a trap stolen by the indians and a number of them had him surrounded but he was relieved by some trappers coming from above who drove the indians off and would have punished

NAMES. Children	Month	Day	Year
Sally Smith was born	Octo.	8	1791
[illegible] Smith " "	Apl.	6	1793
Ralph Smith " "	Sept.	11	1794
Betsey Smith " "	Feby.	12	1796
Eunice Smith " "	Octo.	7	1797
Jedediah S. Smith " "	Jany.	6	1799
Maria Smith " "	Jany.	6th	1802
Almira Smith " "	Aug.	11th	1804
[illegible] Smith " "	Feby.	9th	1807
Austin Smith " "	June	4	1808
Peter Smith " "	March	1[illegible]	1810
Ira G. Smith " "	Octo.	18	1811
Paddock Smith " "	Jany.	19	1812
Nelson J. Smith " "	Dec.	22	1814

Dressing case carried by Jedediah Smith on the Santa Fé trail. **Photo courtesy Historical Collection, Security-First National Bank of Los Angeles.**

Bible record, dated 1834, in the possession of Mrs. C. F. Calhoun, Los Angeles, descendant of Solòmon A. Simon and Eunice Smith; evidently copied from an older record and in error on the birth date of Paddock Smith. A similar record is preserved in a bible owned by Miss May Davis, Mt. Pleasant, Ia., Miss Davis is a descendant of Betsey Smith. **Photo courtesy Historical Collection, Security-First National Bank of Los Angeles.**

them had it not been for some miry ground over which they retreated.

20th I went with the Trappers within a mile of the place where I struck the river on the last Ap[l]. Above that there was no Beaver sign but considerable from the camp up to that place. I saw some indians on the opposite side of the river but they ran off. The river was quite rapid and the rushing of the water brought fresh to my remembrance the cascades of Mt. Joseph and the unpleasant times I had passed there when surrounded by the snow which continued falling. My horses freezing, my men discouraged and our utmost exertion necessary to keep from freezing to death. I then thought of the vanity of riches and of all those objects that lead men in the perilous paths of adventure. It seems that in times like those men return to reason and make the true estimate of things. they throw by the gaudy baubles of ambition and embrace the solid comforts of domestic life. But a few days of rest makes the sailor forget the storm and embark again on the perilous Ocean and I suppose that like him I would soon become weary of rest.

21st Nothing material occurred the weather rainy. 9 Beaver were taken.

22nd I moved across the River.[116] My goods were ferried over in a skin canoe, and the horses swam. The Skins of my 3 Canoes had been carried along for several days and were ready for us at any time by making a frame. After crossing the River I moved ¾ of a mile and encamped on a slou of the River my camp being on an island. It was so late when I encamped that but few traps were set. The afternoon was rainy. One trap was lost, another broken and 11 taken by Reed & Pompare, leaving me but 32.

23[d] I took all the trappers and went down to the forks and up the Main River but soon found slous so deep as to be swimming. I then turned back and endeavored to head them but found it so Muddy that the horses could not travel. Found some indian Lodges deserted but on searching found two squaws, one too old to run away and the other blind. They were trembling with fear and made signs for us to go away. I gave them an awl and some pieces of flannel that I had in my Shot Bag at the same time I gave them some fish that the men found in one of the Lodges. This seemed to satisfy them and they altered their tone so much as to invite me to sit down.

All appearances for progress were unfavorable for as far as I could see up the Main River The flag Ponds & Lakes extended. I hardly knew what course to pursue, for it was impossible to travel North and useless to travel up Wild River on which I was encamped for there was no Beaver in that direction. At 1 O Clock the weather became clear with a north wind. I crossed over the slou by the means of my skin Canoes and a raft made of logs. My raft was formed of some logs that appeared to have been hewn many years since and used for the same purpose to which I applied them. In none of the indian lodges of the vicinity could I see any thing like axes.

24th I concluded I would move up Wild River. But did not go far as I was obliged to raft two slous in 40 yards.

25th E 3 miles up Wild River.[100] Mr. Rodgers and myself went hunting and killed an antelope. We could go no distance from the River on account of the Mud which made the country quite impassable for horses. This was the more surprising as the country was timbered and the soil gravelly. In the evening some of the men found setting for their traps.

26th I went hunting and killed a goose and an Antelope. Two of my trappers, [Toussaint] Mareshall and [John] Turner were up 3 or 4 miles from camp and seeing some indians around their traps who would not come to them but attempted to run off they fired at them and Turner killed one and Mareshall wounded another. I was extremely sorry for the occurrence and reprimanded them severely for their impolitic conduct. To prevent the recurrence of such an act the only remedy in my power was to forbid them the privilege of setting traps, for I could not always have the trappers under my eye.

27th fine weather but still so muddy that I was afraid to try the country North. I went down the river a few miles in doing which I fell in with an indian who could not handily get away and coaxed him to camp. I made him a few presents and sent him off.

28th remained at the same camp.

29th As the only chance was to go down the river I moved down to my old camp and endeavored to go further but found it useless.

March 1st 1828 I went in company with the trappers down to the confluence of Wild River and the Buenaventura which was about 2 Miles from camp. The Buenaventura still continued about 300 yards wide and came from the North maintaining the appearance of which I have before spoken. The Mountain on each side about 30 Miles distant. In going down Wild River we came suddenly on an indian lodge. Its inhabitants immediately fl[e]d. Some plunged into the river and some took a raft while some squaws ran down the bank of the stream.

We galloped after them and overtook one who appered very much frightened and pacified her in the usual manner by making her some presents. I then went on to the place where I had seen one fall down. She was still laying there and apparently lifeless. She was 10 or 11 years old. I got down from my horse and found that she was in fact dead. Could it be possible, thought I, that we who called ourselves Christians were such frightful objects as to scare poor savages to death. But I had little time for meditation for it was necessary that I should provide for the wants of my party and endeavor to extricate myself from the ebarrassing situation in which I was placed. I therefore to convince the friends of the poor girl of my regret for what had been done covered her Body with a Blanket and left some trifles near by and in commemoration of the singular wildness of those indians and the novel occurrence that made it appear so forcibly I named the River on which it happened Wild River. To this River I had before that time applied a different name.

I found so many Ponds Lakes and Slous along the Buenaventura immediately above the mouth of Wild River that I thought the country impassible. On our return towards Camp in crossing a Slou which was swimming deep 2 horses were drowned. On one was six traps and on the other four. The stream was not more than 20 yards in width but a strong current and filled with trees. We went home with the intention of fixing one of the Skin Canoes and coming down the next day to search for them.

March 2nd I sent some of the men to look for the drowned horses and went myself to see if it was possible to kill a Deer. I killed one and wounded two others and on coming in the men let me know that they could [not][116a] find but one of the drowned horse which luckily was that on which was the six traps. I then went in search myself but was unsuccessful. The loss of traps in that country and in those times was much regreted as I had but verry few and there was no chance to procure more. During our stay at this encampment the trappers brought in 3 or 4 Beaver each day. But one meal of meat ahead.

March 3d N 12 Miles finding that the water had somewhat dried out of the ground I determined to make one more attempt to proceed. I found the mud verry deep but not as bad as I had expected. There was an indian trail leading through the most difficult part of the way which served as a guide. We passed 30 or 40 indian lodges but the indians as usual all ran off. I encamped at the head of a flag Lake and at some lodges which the owners left for our accomodation. After encg I saw a band of elk and taking with me 3 Men we killed two Barren does which were in good order for the season. The weather fine and warm. Muskitoes troublesome.

March 4th N 11 Miles and encamp on a creek 20 yards wide running SW.[117] As it was verry Brushy I called it Brush Creek. found some Beaver sign and had traps set. Saw a good many Elk and passed on the bank of the Main river which continues to run North and South several indian lodges thatched with grass.

March 5th Mr. Rodgers went hunting and I went with the trappers. We crossed over the creek on a tree which had been felled for that purpose and went to the River but found that the main river had turned to the left some miles below. On this fork bearing NNE[118] found some Beaver sign and had my traps all set. Mr. Rodgers saw a good many Elk but as the country was not favorable for approaching them he did not kill any.

March 6th I sent my trappers over to their traps and I moved on up the creek to find a place to cross. I threw a tree across but it would not answer the purpose and finding the banks high and some Slous I abandoned the idea of crossing at that place and returning to camp I had a skin canoe made with the intention of going down the creek to cross. My men who were off trapping came and encamped opposite.

7th March I moved down a mile and crossed by the help of the Skin Canoe the horses swimming. All got over safe but it was too late to move. I went down to where the trappers were and carried them some Blankets. They had killed a large Brown Bear which was in good order and were of course feasting. Yes, I repeat it, feasting, for the hunter of the Buenaventura Valley at the distance of 2000 miles from his home may enjoy and be thankful for such Blessings as heaven may throw in his way.

8th N N E 7 Miles I was under the necessity of travelling verry crooked to avoid the mud encamped on the smaller River to which I had not at that time applied a name. Opposite to my camp was an indian village and not far below one or two more. Their Lodges were built like those of the Pawnees. After we had encamped several of them came and sat on the bank opposite talking but in a language which I did not understand. On this river I found a plenty of Beaver sign. 13 were caught the first setting. Mr. Rodgers killed a Brown Bear and wounded another.

March 9th Early in the Morning Mr. Rodgers went after the wounded Bear in company with John Hanna. In a short time Hanna came running in and said that they had found the Bear in a verry bad thicket. That he suddenly rose from his bed and rushed on them. Mr. Rodgers fired a moment before the Bear caught him. After biting him in several places he went off, but Hanna shot him again, when he returned, caught Mr. Rodgers and gave him several additional wounds. I went out with a horse to bring him in and found him verry badly wounded being severely cut in [many][118a] 10 or 12 different places. I washed his wounds and dressed them with plasters of soap and sugar.

The indians came as they had done the day before and sat on the bank of the River. I prevailed on several of them to come over and made them presents of Beads, pieces of Flannel and some Meat. They were entirely naked. The game of the Country was Bear Elk Black tailed Deer Antelope Large and small Wolves Beaver Otter and Raccoon. The Birds were Swan Geese Crane Heron Loons Brant Many kinds of Ducks Indian Hens. Some small birds but they were not plenty. The birds of Prey were Buzzards Crows Ravens Magpies &c. The trappers took 9 Beaver.

10th March Mr. Rodgers wounds were verry painful. I dressed them frequently with cold water and salve of Sugar and Soap. The indians came across the river again bringing me presents of several Bunchs of feathers worn on the head. 11 Beaver were taken.

11th March. The indians came to the opposite bank of the River as they had before done. I invited them over and made them some presents. As I was intending to remain at that place for some time I concluded to go to their village. With this intention I took some small presents and was ferried over the river by the indians on their log rafts.

Arrived at the village I was seated on a mat in a vacancy apparently left for Public use and comenced business by giving my presents. The principal characters took them for distribution in doing which they were verry exact giving to some one and to others two or three Beads as their respective merits might claim. In making their division they did not speak loud but whispered among themselves.

After this business was finished they I suppose felt under obligation to make some presents in return and commenced bringing me fishing nets and dishes but I returned them expressing by signs my satisfaction and my desire to return. When I came away they endeavored to cry as a demonstration of their sorrow for my departure. The village consisted of about 50 Lodges. I saw nothing among them which had any appearance of having come from a civilized country. They were generally naked but a few of them had feather robes and dresses made of net work. The dress of the women consisted of a belt around the waist to which was attached two bunches of bark or flags one hanging down before and the other behind in the form of a fringe.

These indians smoke in wooden pipes and in common with the most of the indians of this valley they wear their hair not more than 5 or 6 inches in lenght. The entrance to their houses is by a low passage covered with dirt through which they are obliged to creep on their hands and knees. 20 Beaver were taken and as I had but 28 traps I considered it great trapping.

12th March The men were trapping several miles above

camp towards the mountain. they told me that the river forked about a mile above camp and that the fork on which they were trapping was clear and had some rapids.[119] In the course of the day some indians came to camp for meat as usual. 8 Beaver taken.

13th March Remained at the same camp. In the vicinity were a good many Black tailed Deer and I improved every opportunity to dry the best of the meat. I sent men up the river to the Mountain but they found verry little Beaver sign higher up than where the trappers had been setting. 13 Beaver taken.

14th March I made my calculations for crossing the fork which came from the East on the Morrow. To this river I gave the Indian name Hen-neet.[120] The weather still continued fine and Mr. Rodgers wounds in such a situation as to make it impossible that I should move any great distance in the day.

15th March I went with the trappers across the Hen neet and directed them to encamp near where the[y] would [encamp][120a] set their traps. I recrossed to camp. A considerable number of indians crossed the River a short distance below. 14 Beaver taken.

16th March Moved N E about 1 mile up the river and crossed over above the forks without any difficulty by the help of my skin canoe in which my goods were carried over, the horses swimming. The indians near my camp still continue friendly and were singing when I left them. 12 Beaver taken.

17th March I went with the trappers 8 or 10 Miles up the River which came from the North and united with the Hen neet near my camp. To this River I gave the name Ya-loo[121] which was the name I applied to the indians of the village last visited. We found but little Beaver sign as far as we went up the river. I passed two indian villages of 20 or 25 dirt lodges each the inhabitants were much alarmed at our approach but after some time I prevailed on them to come to me and take some presents.

During my absence there was a considerable alarm in camp by the appearance of several hundred indians on the opposite bank of the River but the alarm subsided when they passed on up the river and in a short time returned loaded with Acorns from some caches they had in that direction. 9 Beaver taken.

18th March I sent the trappers to remove their traps from the Yaloo to a branch of the Hen-neet on which they had found some Beaver sign. 9 Beaver taken.

19th March As there was no chance for trapping on the Yaloo and some little on the Henneet I moved N E 5 miles towards the foot of the Mountain. I was induced to do this more especially as Mr. Rodgers was not in a situation to make a great days travel which would be necessary in traveling up the Yaloo for it was in all probability some distance to Beaver in that direction. After encamping I went [up the river][121a] North 5 or 6 miles to a Creek 20 yards wide running west called from the Color of its bank Red Bank Creek.[122] It was fordable but I found verry little beaver sign in it. The country generally as the dry season advanced became dry and firm and apparently fit for cultivation, presenting a verry different appearance from that of a month back when it was almost impossible to travel in any direction.

In the course of the days travel I saw some Antelope and the sign of Elk and fell in with 2 indians and a squaw on the plain. I found they were attending [their][122a] some nets set for the purpose of catching Brant. From where they stood cords extended 2 or 300 yards to the nets and there we observed several Brant.

While I was making the indians presents of some Beads the men said they would creep up and kill some. They made several shots without success when I told them I thought they were deceived. They said not for they had seen them move and one said his gun must be crooked but that he would try them again. He did so and I was convinced that I was right in the supposition that they were decoys and on examination found them so complete that the deception could not be detected except in verry near approach. The nets were about 20 feet long and 6 feet wide and arranged much like the common pigeon net. There was three of them all to be sprung at once by the same line. 9 Beaver taken.

20th March On account of the wounds of Mr. Rodgers I was obliged to remain in the same camp. The weather still continued fine, 9 Beaver taken. Some indians came near camp I went to them and gave them some presents of Beads and some Meat with which they appeared much pleased.

21st March N W 7 Miles crossing the several channels of Red Bank Creek and encamp within 3 Miles of the Yaloo. After encamping I went with the trappers down to the river where they set their traps. The indians were numerous and in one place I came uppon them before they had an opportunity to run off and gave them some Beads according to my common custom. The squaws had their baskets filled with young Pea vine and from what I could observe I think their principal supports consi[s]ts of Acorns, Grass, Pea vines, Roots and what few fish and water fowl they are able to take.

If Missionaries could be useful in Civilizing and Christianizing any indians in the World their efforts should be turned towards this valley. The indians are numerous honest and peaceable in their dispositions. They live in a country where the soil is good and the climate pleasant with the exception of 2 or 3 months in the winter when there is too much rain. There is seldom any frost and I have seen snow but once in the valley of the Buenaventura.

A great many of these indians appear to be the lowest intermediate link between man and the Brute creation. In the construction of houses they are either from indolence or from a deficiency of genius inferior to the Beaver and many of them live without any thing in the shape of a house and rise from their bed of earth in the morning like the animals around them and rove about in search of food. If they find it it is well if not they go hungry. But hunger does not teach them providence. Each day is left to take care of itself.[123] . . degraded ignorant as these indians must be and miserable as the life appears which they lead it is made more apparent by a contrast with the country in which they are placed a country one would think rather calculated to expand than restrain the energies of man a country where the creator has scattered a more than ordinary Share of his bounties. [Another observation I made among these indians][123a]

22nd March Lay by on account of the wounds of Mr. Rodgers. Some rain during the night. Nothing material occured. 12 Beaver taken.

23d N W 6 Miles and encamp on the Yaloo River. At 12 O Clock it commenced raining and continued until 3 when it cleared off with a west wind. The trappers who came directly up the River passed 3 indian villages. They found the river somewhat rapid and but little appearance of Beaver.

24th March The party remained in camp. I went with several of the men a North East course to the foot of the Mountain. Betwe[e] camp and the Mountain on the North side of the river were two indian villages and some Beaver. Some indians visited camp in my absence and Mr. Rodgers gave them some Beads and some Meat. 17 Beaver taken.

25th March W N W 3 Miles to the place I had selected for crossing the river. 17 Beaver taken.

26th March The Ya loo at that place was about 100 yards wide and strong current but as I found good bars[124] on both sides of the river I had no difficulty in swimming my horses and my goods were carried over in a Skin Canoe made for the purpose. The indians were close to camp on both sides of the river but I did not allow them to camp this being with me a general rule.

I went to them and gave them some presents and one of them was so bold as to venture into camp but soon left it at my request. But few traps were set. 10 Beaver taken.

27th March I had my horses caught early but one of them ran off and swam the river. I got an Ind^{n} raft which was a mile above and went over after him and drove him back. At 12 O Clock we moved off W N W 12 Miles over a level Prairae crossing some small muddy Creeks and encamp on a Creek[125] 30 yds wide running SW Deep and Muddy with some timber on its banks but very little Beaver sign. I had a tree thrown across at a narrow place to form a foot Bridge that I might be in readiness to cross early in the morning as I was desirous of proceeding to the Buenaventura River which appeared about six Miles distant.[126]

28th March 7 Miles W N W and encamp on the Buenaventura. In the course of the day I crossed two muddy Slous of the River. The Buenaventura at that place was about 200 yards wide Deep and forcible current. Its general course South and its banks fringed with timber principally Cotton wood and Sycamore and when the banks were somewhat higher Oak. Far off to the north verry high Peaks of the Mountain were seen covered with snow. The valley at that place was apparently about 50 Miles in width. The Mountain to the west on towards the coast not high but rugged and some snow. On the East the Mountain was high timbered and its upper region covered with snow. In the course of the day I saw some Elk and the trappers killed two they were in good order. There was not much Beaver sign about the river its banks were too sandy, But a short distance back were Lakes and ponds in which were found some Beaver.

29th March N 6 Miles and encamp on the river. I was obliged to cross many Slous of the River that were verry miry and passed great numbers of indians who were engaged in digging Roots. I succeeded in giving to them some presents.

they were small in size and apparently verry poor and miserable. The most of them had little Rabit Skin Robes. 11 Beaver taken.

30th March The Party remained in camp and I went up the river with one man to examine the country. About 1 Mile above camp a creek came in 20 yards called Pen-min wide deep and Muddy.[127] Along its banks were many dirt Lodges having the entrance at the top. As we passed along the little children reminded me of young wolves or Prairae dogs. They would sit and gaze at us until we approached near to them when they would drop down into their holes. Some of the indians appeared much frightened as we came in sight while others scarcely quit their Work (digging roots) to look at us.

At that place I saw a few indians who wore their hair long [but.][127a] The women dressed like the last described except perhaps that the scanty apron was there sometimes made of Deer Skin instead of bark or flags. I came to one place where there was several lodges together the women cryed and the men harrangued me on my approach but I soon pacified them. On my return to camp I got my horse mired and we were obliged to draw him through the mud for two hundred yards 3 of the indians assisting us. On my arrival at camp I found the indians had been there all day or as near as I allow them to come. The indians of this vicinity were all pleased to get the least morsel of meat.

31st March North 8 Miles. To make this distance my route was quite circuitous being obliged to travel much out of my way in order to find a suitable place to cross Pen-min Creek and the mud beyond. We passed many indians and some of them went with us to the place of encamping. Just before encamping I discovered 2 Bear and 3 of us approached them and killed both. They were neither large nor fat. 4 of the trappers did not come in to camp. 14 Beaver taken.

April 1st The trappers all came in one trap lost by Beaver.

In the evening several of us went out hunting for there was considerable sign of Bear Deer Elk and Antelope in the neighborhood. Mr. [Martin] McCoy and J[oseph] Palmer killed a large Grizly Bear in tolerable order and on opening him found nearly in the center of the lights a stone Arrow head together with about 3 inches of the Shaft attached to it. The men brought that part of the lights containing the arrow into camp. The wound appeared perfectly healed and closed around the arrow. 3 indians who came with us to camp were busily employed on the share of Meat alloted to them and on the entrails of the Bear. They filled themselves so completely that they were puffed up like Bladders. One of those indians had a spear with a stone head like that of an Arrow but 5 or 6 times as large. The handle was about 6 feet Long.

2nd April At the same camp. Several indians visited me to whom I gave some small presents and some meat for these indians were well pleased whenever they could [get] the least morsel of meat. Two of the indians who came with us to camp still remained. 17 Beaver taken. For 4 Days past the Cranes and Brant had been on their passage North in great numbers. The Geese had principally gone before.

3d April Remained at the same camp. The weather warm and at night the Musquitoes troublesome. My two indians still with me. 13 Beaver taken.

4th April The trappers had been 4 or 5 miles up the river. Beaver plenty but the numerous Slous interfere much with trapping. The indians were verry numerous and friendly. 18 Beaver taken.

5th April W N W 7 Miles Turned out from the river and 5 miles from camp crossed a Creek 20 yards wide running West. Rapid but fordable. I called it Black Sand Creek.[128] My encampment was on the River bank. Many indians came as near the camp as I would permit and sat down. I gave them some presents. They were naked but had not the miserable appearance of those below. They were under the impression that the horses could understand them and when they were passing they talked to them and made signs as to the men.

6th Remained at the same camp. At 8 O Clock about 100 indians visited us they were generally naked but a few of them had rabbit Skin Robes. They were about 5 feet 10 inches in heighth rather light complexion round featured, wide mouths, and short hair. They brought with them no weapons but had gen^ly in their hands a bush of green leaves. I met them according to my usual custom about 80 yards from camp and invited them to sit down. One who seemed tolerably intelligent showed me the principal men. I gave them some cotton shirting Beads Awls, and Tobacco. They were apparently fond of smoking. Their pipes were long strait and made of wood. Those indians were frequently where my men were setting their traps But did no further damage than springing a few of them. A river about 70 yards in width entered the Buenaventura on the west side 12 Miles below Camp. Its water had a verry Claey appearance. I called it Pom-che-le-ne.[129] Beaver taken.

7th April W N W 8 Miles. At 2 Miles from camp crossed a creek[130] 30 yards wide rapid and stoney Bottom running SW and having some Beaver sign. 3 Miles farther struck a creek[131] same size and running S. W but so deep that I was obliged to follow it up 3 Miles to find a ford at which place I encamped. In the vicinity was considerable appearance of game and particularly bear. In the evening we shot several Bear and they ran into thickets that were convenient. Several of us followed one that was Badly wounded into a thicket. We went on foot because the thicket was too close to admit a Man on horse back.

As we advanced I saw one and shot him in the head when he immediately [tumbled][131a] fell—Apparently dead. I went in to bring him out without loading my gun and when I arrived within 4 yards of the place where the Bear lay the man that was following me close behind spoke and said "He is alive". I told him in answer that he was certainly dead and was observing the one I had shot so intently that I did not see one that lay close by his side which was the one the man behind me had reference to. At that moment the Bear sprang towards us with open mouth and making no pleasant noise.

Fortunately the thhicket was close on the bank of the creek and the second spring I plunged head foremost into the water. The Bear ran over the man next to me and made a furious rush on the third man Joseph Lapoint. But Lapoint had by good fortune a Bayonet fixed on his gun and as the Bear came in he gave him a severe wound in the neck which induced him to change his course and run into another thicket close at hand. We followed him there and found another in company with him. One of them we killed and the other went off Badly wounded.

I then went on horse Back with two men to look for another that was wounded. I rode up close to the thicket in which I supposed him to be and rode round it several times halloeing but without making any discovery. I rode up for a last look when the Bear sprang for the horse. He was so close that the horse could not be got underway before he caught him by the tail. The Horse being strong and much frightened exetered himself so powerfully that he gave the Bear no opportunity to close uppon him and actually drew him 40 or 50 yards before he relinquished his hold.

The Bear did not continue the pursuit but went off and [I] was quite glad to get rid of his company on any terms and returned to camp to feast on the spoils and talk of the incidents of our eventful hunt. 16 Beaver taken.

8th April The party remained at the same camp. I [remained][131b] went up the Creek which I called Grizly Bear Creek to the foot of the first small range of Mountain. The distance was but 1½ Mile and the creek for that distance had a rapid current and stoney bottom. From the top of the mountain which appeared to be a spur of the main range breaking off from it a few miles south of my position I took a view of the country around. On the East Mt. Joseph appeared lower than it had before been having but little snow on such of its summits as were in view. The Main Mountain still ranged nearly North & South and a stream joined the Grizly Bear Creek on the north side. One Bear killed and ten Beaver taken.

9th April At the same camp. Rainy with a south wind.

10th April N W 6 miles. I moved on with the intention of traveling up the Buenaventura but soon found the rocky hills coming in so close to the river as to make it impossible to travel. I went on in advance of the party and ascending a high point took a view of the country and found the river coming from the N E and running apparently for 20 or 30 Miles through ragged rocky hills. The mountain beyond appeared too high to cross at that season of the year or perhaps at any other.

Believing it impossible to travel up the river I turned Back into the valley and encamped on the river with the intention of crossing. For this purpose I set some men at work to make a skin canoe. My Camp seemed in a curve of the Mountain. Mt. Joseph gradually bending to the west appeared in conjunction with the low range on the west side of the river which in its course north joined it to encircle the sources of the Buenaventura. The distance from camp to the main range of Mt. Joseph on the East was about 20 Miles on the N E 30 on the N 25 and the low range on the West about 20 miles. The country East N East and North was hilly rocky and timbered with small Oak and Pine. 10 Beaver taken.

11th April The Canoe being finished I crossed my things over in it and swam the horses. All got over safe with the exception of a colt which was drowned. The trappers found setting for a few traps. 12 Beaver taken.

12th April at the same camp. I went with the trappers down the river to look for Beaver sign but found so little that I did not think it worth setting for. On my way I came suddenly on about 20 indians. The moment they saw me they sprang to their feet and commenced dancing in which they appeared to exert their best energies throwing their bodies into every imaginable position.

I was much surprised at this singular reception and knew not how to consider it. Perhaps it was meant as a charm or a Medicine according to the meaning of the term when applied to indians or it may have been a mark of respect or the accustomed manner of saluting strangers. Be that as it may I soon made them quit and gave them some presents after which they accompanied me as near the camp as I would allow them to go and received some meat which they as well as all the indians of the valley appeared to eat with great relish. 5 Beaver taken.

13th April N W 8 Verry hilly and rough traveling the timber generally scrubby Oak. Some indians came to us on the route we gave them a part of an Antelope which Mr. Rodgers had killed and they left us. My route was in the direction of a Gap of the Mountain through which I intended to pass. I encamped about 12 O Clock to dry my things which were wet by the last rain and stretch some Beaver skins which I had on hand. One of the indians which came to me had some wampum and Beads. They were procured as I supposed from some trapping party of the Hudsons Bay Company which came in that direction from their establishment on the Columbia.

14th April W N W 6 Miles and encamped on a creek 20 yards wide running N E. Some indians who had encamped near me traveled in company. The hill country and some unbroke mules which I had packed prevented my travelling far.

15th April W N W 12 Miles. At 1½ Mile from camp crossed a Creek 15 yards wide running N E. The country verry rough and hilly but fortunately a ridge or divide ran nearly in the direction in which I wished to travel on the top of which I was enabled to move on without much difficulty until nearly night when I turned a little N E and went down into a deep ravine to encamp on the bank of a rapid stream 20 yards wide running S E. I drove the horses under a steep bank next the Creek that I might have a convenient place to catch them.

While catching them I observed an arrow in the neck of a horse and immediately called to the men to tie the horses they had in their hands and spring to their Guns. This was quickly done and several men mounting their horses rode quickly to a point where 10 or 12 indians were throwing their arrows into camp. They ran off and were fired at and two fell but afterwards crawled off. I got a shot at one soon after but he went off leaving much blood behind. The indians were shouting about until night but did not come again within gun shot. In the affray they wounded 9 horses and [two][131c] mules Some Badly and some slightly and in all probability paid for the damage they had done me by the sacrafice of two or three of their lives. In taking out the arrows [points][131d] some of the points were left in. The Creek on which we had encamped had some appearance of Beaver.

16th April West 12 Miles. On account of the roughness of the country I was obliged to turn West. The traveling was exceedingly bad and through a country timbered with some Oak and an abundance of Bastard Cedar. One of the mules and one of the horses both wounded were left behind being unable to travel. I encamped at the foot of the Mountain which was on the West and North West.

The indians had been following us all day and yelling from the high points and after encamping they came quite close to camp. I took several men with me and went within gun shot endeavoring by signs to persuade the indians to come to me being desirous to convince them of my disposition to be friendly. But they had their bows strung and their arrows in their hands and by the violence of their gestures, their constant yelling and their refusal [left no][131e] to come to me left no doubt on my mind of their inclination to be hostile. I therefore in order to intimidate them and prevent them from doing me further injury fired on them. One fell at once and another shortly after and the indians ran off leaving some of their property on the ground. Lest they should return and shoot some of my horses I had a pen made to put them in and had them guarded as I had done the night before.

17th April W N W 10 Miles and then N W 6 Miles. At ½ Miles from camp I crossed a creek 15 yards wide running East. From that place the ascent of the Mountain for 10 Miles was in some places quite steep and timbered with Oak & Pine. Then crossing the ridge of the Mountain where there was some snow and high peaks on the right and left I came to waters that ran to the North West and in a few Miles came to a creek 15 yards wide on which I encamped after travelling down it a short distance. I was apprehensive from the appearance of the country that the stream on which I encamped turned back again into the valley of the Buenaventura.

18th April N 3 Miles. Whilst preparing my horses for starting I sent two men down the creek to see what chance there would be for passing in that direction. It was eleven O Clock before they returned and reported the pass practicable but verry rough. When I moved on I found it necessary to wind among high and steep hills and making but 3 miles encamped on the creek below the mouth of a small creek coming in from the East where there was good grass.

19th April North 6 Miles and West 4 Miles following down the river.[132] The mountain came in close to the river but leaving a tolerable pass along its banks. I encamped where there was a small valley and a creek coming in from the south. At camp the River ran North West. A short distance above camp I saw an Indian lodge and went to it and found an old man a woman and child. The woman and child ran off but the old man staid and I gave him some Beads and tobacco. Soon after encamping some indians showed themselves on the opposite bank of the river and appeared to be creeping up to get a shot at the horses. I went close to them with Arthur Black made friendly signs and invited them to come to me. But they answered by prancing about and making preparations to throw their arrows. I therefore told Black to fire. He did so but without killing any and the indians ran off letting fly their arrows as they went. I shot as they ran but did not kill. 12 or 15 indians soon collected on the other side of the river at the distance of 400 yards and made a fire. Soon after some of them came close enough to throw some arrows near the horses. Some of my men fired at them but without success. About sunset 6 or 8 came and made another fire a short distance from camp. I had 4 horses caught and 3 men and myself gave them a chase. 2 of them were killed and the rest escaped. After this they troubled us no more.

20th April. N N W 8 Miles following down the river I had to cross some high points of the mountain and travel along the side of the hills through thickets of Brush and over steep and [precip][132a] rough masses of rock. The traveling extremely bad was made much more difficult and dangerous by the great number of horses which I had along. In a bad pass the horses all endeavored to avoid being crowded off themselves and therefore rushed against whatever opposed them.

In a struggle of this kind two of my horses were pushed from a precipice into the river and drowned. 5 traps and some of the mens things were lost. I was obliged to encamp where there was but verry little for my horses to eat.

21st April W N W 12 Miles Traveling same as yesterday. I found some grass on the side of the Mountain about 1 mile from the river and encamped. There appeared to be a small valley on the river which turned in its course nearly North and [nearly North][132b] received a branch from the South. Several Smokes were in sight during the day and several of my horses were verry lame from the roughness of the traveling.

22[d] N W 3 Miles and encamped where there was good grass in a small valley on the ——. Passed several indian Lodges the indians themselves were yelling on the hills and some appeared in sight of camp, but when I attempted to go to them they ran off. The river to which I had given the name Smiths river[133] had been gradually increasing in size and at my camp was about 40 yards wide with a strong current and wide sand bars. Its course was N N W. The Mountains of the vicinity were covered with Pine timber and the summit covered with snow.

Just before sunset one of my horse guard came in and told me that there were some indians on the opposite side of the river close at hand. I went with one man to see what they wanted but before I got down they were throwing their arrows at the guard. They were at a distance of 150 yards but their arrows scarcely reached us. I called for some men and went down to the bank of the river and fired several guns wounding one or two of them but killed none dead on the ground. They then ran off yelling and troubled us no more that night. Among those troublesome indians I was obliged to put my horses in a pen every night and have them guarded the fore part of the night but as those indians had but little clothing and the weather in those mountains was cold there was no necessity for continuing the guard during the latter part of the night. Several of my horses were verry lame.

23[d] April I got under way quite early and went down the river about two miles but the mountain came in close to the river so that I was obliged to order the party back to camp as there was no possibility of proceeding without crossing the river. I went with two men to look for a pass. On examination

I soon found the best course would be to cross the river and for that purpose found a ford and returned to camp. No indians were about in sight.

24th North 7 Miles. Early in the morning crossed the river without any material accident and continued down the river the mountain coming in quite close to the river with Brushy thickets and deep ravines. One point of the Mountain over which I was obliged to pass was so exceedingly Rocky and rough that I was four hours in moving one mile. The Rocky hills over which we had to clamber mangled the feet of the horses most terribly. At my camp [were][133a] the steep side hills were covered with Oak and Pine timber and the grass was tolerably good. I observed a kind of tree with which I was before unacquainted. The largest were 1½ feet in diameter and 60 feet high. The limbs smooth and the bark snuff colored. It was at that time in Bloom. Some Europeans who were of my party called it the Red Laurel.

25th My horses were so much fatigued that I remained in camp. Several of us went out to hunt and killed 3 Deer [they ate.][133b] Some of the men found some nooses set to catch Deer. They make a fence of Brush leaving a small aperture over which a cord is extended with a noose sufficiently long to admit the head of a deer. It is of course set in some of the common passes.

26th 5 Miles N W About two Miles down the river and immediately below the mouth of a creek coming in from the West the Mountain closed in to the river which ran in a channel of cleft rocks. I therefore turned up the creek and encamped on the north side where it was 30 yards wide rapid and difficult fording. The traveling rough and rocky being along the abrupt sides of the mountain on which were some Oak Pine and Hemlock timber and tolerable grass. More of my horses and Mules were wounded by the rocks during the days march.

Any persons apprised of the character of the country through which I was traveling might form something of an Idea of the difficulty of traveling with a Band of three hundred horses. After encamping I sent two men to look for the best pass over the mountain which lay on the North. They returned at dark and told me it would not be difficult to ascend to the top of the Mountain but that they could not see far enough to judge of the traveling beyond.

27th As I was not satisfied as to the best route by which to continue my [route][133c] journey and as the grass about my camp was tolerably good I did not move the party but sent 3 men back for a horse that had been left and went myself with one man to view the country. The best traveling I could discover was to steer NW and keep on a range of hills [which was][133d] the divide between the River and the creek which had its rise apparently nearly in the direction to which the river ran. The hunters killed 4 Deer and 2 Grizzly Bear. The men from horse hunting returned having found 2 instead of one.

28th N W 3 Miles ascending the steep side of the mountain and arrived at the top and turning N N W the ground a little descending for three fourths of a mile the snow was three or four feet deep. Leaving in that distance the snow and continuing the same course for 5 Miles over high ridges and through Deep ravines along the sides of abrupt hills and through dense thickets after working hard all day we made but 8 Miles and encamped where I was obliged to make a pen for the horses to keep them from s[t]rag[g]ling off as there was no grass for them. At night it was found that 5 were missing, two of that number being packed one with fur and one with some clothing belonging to the Men. By the help of a good Moon light two of them were found before I went to bed.

29th North 3 Miles As soon [as it was] light I sent the [company] forward and went myself with 3 Men to look for the lost horses. Found them in different places and safe the packs on the Mules having remained on during the night without turning. I got to camp about sunset and found good grass.

30th N [W][133e] 1½ mile with the intention of going to the river but I found the deep ravines impassable and the river yet washing the base of high hills. I there[fore] retraced my steps to a place where I had seen good grass and encamped sending men off at the same time to see if there was any possibility of passing back from the river. When the[y] returned they told me they thought it passible although the traveling would be bad.

May 1st 1828 North West 3 Miles I went but little beyond where the men had gone the day before when I found the traveling so bad that I was obliged to encamp and send on again to search for a pass. At my camp there was verry little grass. The men returned and reported the traveling extremely bad for about three Miles after which there was plenty of good grass. I went hunting with several men we killed [several][133f] one Deer which was quite in time for our dried meat was nearly exhausted. Rain with some snow during Most of the day and following night.

May 2nd North 2 Miles. The road most terrible down steep hills which were extremely Rocky and Brushy. On the side of the mountain were some remarkably handsome hemlocks, the largest I had ever seen. Beside Hemlock was Pine and some Oak. On the point of a ridge on which I encamped was some good grass. 4 Deer were killed.

May 3d 1 Mile North. I first made an attempt to move down towards the river but found it impracticable. I therefore returned to camp and moved north 1 Mile over traveling like that to which I had now become accustomed. I encamped a mile from the river on a ridge which produced plenty of grass and Oak timber. Opposite my camp a large stream entered Smiths River from the East.[134] It appeared even larger than the stream on which I had been traveling. One Mule lost. After encamping the hunters went out and killed two deer.

May 4 I was obliged to lay by in consequence of the lameness of my horses. I had my Beaver skins dried and sent men back on the trail to look for horses. The hunters killed 8 Deer and the Meat was cut and dried.

5th May At the same camp, some of my horses being unable to travel. I had my horses brought up and counted and found that there was ten or twelve not to be accounted for. I therefore took one man and went back on the trail intending to go [back][134a] to the 4th encampment directing Mr. Rodgers that in case [I was back][134b] I did not return to start early on the following morning. I found two horses and got two miles on my way back.

May 6th When I got to the party in the morning they were 3 Miles on their way traveling north. For that distance the road was tolerable being near the river. The Mountain came in near the river but was not so abrupt as it had been nor so high, particularly on the west side of the river. Passed several indian lodges and encamped opposite to one. Their Lodges were built differently from any I had before seen. They were 10 or 12 feet square, the sides 3 feet high and the roof shaped like a house. They were [shaped][134c] built of split pine plank with 2 or 3 small holes to creep in at. About ½ Mile above camp a creek entered on the west side 20 yards wide. Rapid current.

After camping a canoe came down the river with a good many Deer skins on board. I made signs for them to come to [me] but they would not. 2 or 3 indians passed down on the opposite side of the river. I endeavored to persuade them to come over but did not succeed.

May 7th North 4 Miles then North West 5 Miles following the river as close as the traveling would permit. Passed through thickets and over two verry high rocky hills from the last of which the country had a much more promising appearance. Lost several Mules and horses in the course of the day but found them all again. Several indians came to camp in my absence.

They appeared friendly and made signs that they wished to trade Deer Skins for Axes & knives. Indian trails were becoming large and lodges of the kind mentioned more plenty than in the country through which we had for some time been traveling. I saw several places in the course of the day where there had been axes used. Judging from the size of the river and the appearance of the country I suppose the river had in the course of the days travel received a tributary from the East as large or larger than itself.

8th 2 Miles N W In the morning several indians came to camp different from the indians I had before seen in the country, particularly in their dress and in the length of their hair which was long while nearly all the indians of the Buenaventura valley and the country generally I have distinguished by the appellation of short haired indians. These indians were clothed in [skin][134d] Deer Skins Dressed with the hair on. The lower part of the body was left naked. Some of them had Mockasins. Their lodges were tolerably numerous and they had a few good canoes.

Soon after starting a horse ran off and detained me so long that I did travel but two miles before encamping. Two of my horses were found dead when we caught up to move on, poisoned as I supposed by eating some poisonous weed.

9th N W 6 Miles Following the river 3 Miles but it turning more to the North and the indians informing me by signs that it was Rocky along the bank of the river. I turned N W following a ridge which was in that direction and encamped 2 or 3 Miles from the river on a creek. 3 horses lost. An abundance of Elk and some deer sign. One fine Elk killed.

10th[135] N W 5 Miles. To make this distance I traveled as much as ten Miles first attempting to move in towards the River with the intention of traveling along its bank but this I found impracticable and turned back on to the ridge and moved N West untill night over hills rocky and steep and through thickets and deep ravines to a small creek where I encamped without any grass for my horses and was therefore obliged to make a pen for them. On examination I found several were missing, among the rest two that were packed.

May 11th N W 1 Mile. I went up a verry steep hill and finding grass encamped and sent 4 Men back to look for the lost horses and a gun which had been lost at the same time. The men returned in the evening having found 13 horses. There was three yet missing and the gun was not found. The hunters killed three Deer.

12th May I remained at the same camp and sent back two men to look for the lost horses. They found one but could not drive it to camp, therefore they were abandoned.

13th 4 Miles N W I had flattered myself that I was nearly over the bad traveling But I found this day of the old kind. The course of the river was N N W and I made an attempt to go down and travel along its banks but did not succeed and was obliged to wind about among the hills and mountains.

May 14th 2 Miles north. I made another attempt to get in to the river but the rocks obliged me to take to the hills again. Crossing a deep rocky ravine I found greater obstacles than I had before encountered in that rough country. I worked with all my men hard during the day and at night had made but two miles. Two of my horses were dashed in pieces from the precipices and many others terribly mangled. Some of my packs I was forced to leave in the ravine all night with two men to watch them.

May 15th I went back with several men to fetch in the packs and horses left in the ravine and worked hard until 3 O Clock before we got them all to camp. Some men out hunting killed 5 Deer.

May 16th My horses being verry lame I thought it most prudent to remain in camp for a time. I sent two men to examine the country ahead. Some indians came to visit us at camp bringing with them some Lamprey Eels and Roots for trade. I gave them some presents purchased their fish and a Beaver skin which one of them had. They had Beads Wampum & knives. I endeavored to make them understand by signs the direction in which I wished to travel and to ascertain something of the character of the country but they could not understand me. They appeared very friendly and I allowed them the privilege of camp. Some of their squaws was with them.

May 17th Remaining in camp at 9 O Clock the 2 Men sent out to reconnoiter returned. They told me that the country to the West was tolerable [and] that the Ocean was not more than 15 or 18 Miles distant. I determined therefore as traveling along the river was so bad to move towards the coast.

May 18th West 3 Miles along a ridge somewhat thickety and encamped in a small prairae of good grass. In the course of this short days travel two horses gave out. My men were almost as weak as the horses for the poor venison of the country contained verry little nourishment.

19th May West 6 Miles principally along a ridge brushy and timbered with Hemlock Pine & Cedar. Some of the Cedar's were the noblest trees I had ever seen being 12 or 15 feet in diameter tall [and][135a] straight & handsome. I encamped in a prairae with the Ocean in sight. 6 Elk were killed two of them in tolerable order. Counting my horses I found that three were missing. 4 indians that followed us on the trail came up and encamped with us.

May 20th I remained in camp to give my lame horses an opportunity to recruit and dry meat. I sent two men to look for lost horses and Mr. Rodgers & Mr. Virgin to examine the country towards the Coast. Several indians visited the camp in the course of the day. At night Mr. Rodgers and Mr. Virgin returned. They had found the traveling along the coast verry bad. The hills heavily timbered and brushy coming in close to a rocky shore. In their excursion Mr. Rodgers had left

Mr. Virgin a short distance with the horses to get a shot at some Elk. Shortly after he was gone some indians raised the yell and at the same time let fly their arrows at Mr. Virgin and the horses. In return he shot one of them down and calling for Mr. Rodgers at the same time they ran off having wounded one of the horses verry badly but not to prevent bringing him in to camp. The two men horse hunting did not return.

21st May Rainy with a verry heavy fog. The horse hunters returned with one horse. No indians visited camp.

May 22d I had my horses caught up early but just as I was ready for starting it commenced raining and made it impossible to travel for the dense fogs quite common to this coast would prevent me from avoiding the deep ravines and precipices that everywhere came across my way. Among the animals I observed in the country was Elk, Black tailed Deer & Black Bear all of them plenty. Some Raccoons, Large and small wolves, Foxes, Wild Cats, Grey & striped squirrels. The Birds are Large & small Buzards, Crows, Ducks, Ravens, several kinds of hawks, Eagles and a few small birds among which are Robbins & Humming Birds.

23d May From the information I had obtained of the nature of the country on the coast I was convinced of the necessity of retracing my steps and making the attempt to cross Smiths River where I left it.[136] I previously made another unsuccessful attempt to find a pasage along or near the coast and early in the morning taking a man with me I endeavored to find a ridge by which I might pass to the river without following the trail by which I had come out but my efforts were unavailing and I returned to the party and moved back on the trail and encamped at the same place as when I came out. during the past night a north wind cleared the sky from clouds and left the weather fine.

24th May Back to the river and down it [to][136a] 1 mile below my old encampment. Finding [my][136b] no grass on the river I encamped about ¼ of a mile from the west bank. A little below my camp and on the opposite side of the River was an indian village where there was some Canoes. I went down to the River and calling to the indians some of them came over and went with me to camp. I gave them some Beads and made them understand that it was my intention to cross the river the next day. One horse lost and two men sent back to find it. During most of the day a heavy fog. The men returned with the lost horse.

May [2] 5th I packed up early & went down to the river. The indians appeared suspicious that I had some evil intentions and made signs for me to go off. However after a long time I prevailed on one of them to come over with a Canoe. I soon convinced him that I only wished to cross the river and promised to pay him if he would bring 3 or 4 canoes and carry my things over. No sooner were the indians satisfied as to my designs than they brought over their canoes and soon my things were all taken across.[137]

In swimming my horses some of them fell too far down and had hard work to get out and others returned and went out on the same side. I went across with some indians and two men although it was raining and soon found 12 horses & mules hudled together on the bank. We drove them in and they swam over. At the same time we found one horse drowned. When we were all over I sent two men down the river to view the country. during the afternoon the indians visited camp in considerable numbers. They brought with them a few Lamprey Eels & I got of them a piece of salmon, 2 Beaver & an Otter skin. They stole a trap. The explorers returned at night telling me that the country down the river was tolerably rough.

May 26th. North 6 Miles. I moved down the river 2 Miles and then struck east on to high range of hills winding along their summit for several miles and then turning west again in consequence of a creek that made a deep ravine. I encamped about 2 miles from the river having traveled about ten miles to make the six above mentioned. 3 Deer Killed 5 having been killed the day before.

27th May W N W 3 miles and encamped on a creek 30 yards wide running west. My camp was in a small bottom of grass just above the confluence of the Creek and river. The desent to the creek was down a hill long & steep & thick with brush. One horse left and five or 6 lost. I sent two men back for them. Several indians inoffensive in appearance and without arms visited camp.

May 28th North East 7 Miles In consequence of the hills which came in close and precipitous to the river I was obliged to ascend on to a range of hills and follow along their summits which was verry difficult particularly as a dense fog rendered it almost impossible to select the best route. I encamped where there was verry little grass and near where the Mountain made a rapid descent to the north rough & ragged with rocks. I went to the brink of the hill and when the fog cleared away for a moment I could see the country to the North extremely Mountainous along the shore of the Ocean those Mountains somewhat lower. From all appearances I came to the conclusion that I must move in again towards the coast.

May 29th. 1 Mile back on the trail. I attempted to move towards the river but the fog closed around me so thick that I could not see how to travel and finding myself among thickets & Deep ravines I was obliged to stop and send off men to search for a pass. About 2 O Clock it cleared off and I was enabled to see the country around me. The general course of the river was West as far as its entrance. In places are small prairaes along the bank and in others the Mountain closes in to the water.

May 30th 2½ Miles N. W. In the morning it was quite clear on the mountain while the river the deep ravine and the Ocean were hid from view by a dense white Cloud. My route was down a steep hill in [to] the valley of a creek where I encamped in a small prairae of good grass ½ mile from the river. I [went] back with some men to look for some horses and one load that was lost in the descent of the hill and found 7 horses and the load with the exception of one trap. 2 Elk were killed. In the evening it became again foggy.

May 31st In the morning it commenced raining and continued during the day. I therefore did not move camp. In the vicinity I saw a bush resembling a common brier in appearance only somewhat larger. Its fruit was like a raspberry in taste and shape but larger. They were ripe at that time and some were yellow and others red. 2 indians came to camp remained all night and the rain still continued.

June 1st 1828 West 3 Miles and encamped ¼ of a Mile from the river in a small prairae where there was some grass. In the course of the day we traveled through dense thickets and timber and up and down two steep hills made verry bad by the rain of the past day and night and its continuation during the day. Several horses with their loads were lost.[138]

June 2nd Remained at the same camp. At 10 O Clock the sky became [clear?]. Several men sent off for the lost horses returned having found all that I had missed. Two harmless and inoffensive indians visited camp without any arms.

June 3d West 2 Miles. Moving along a ridge passing through a close thicket and down the point of the ridge into the river bottom. I encamped where I was stoped by swamps and muddy ground at the distance of half a mile from the river and where there was hardly any grass for my horses. The tide came up in the river opposite my encampment.

June 4th North 1 Mile. Whilst the party were preparing I went ahead looking [for] a route to pass around the swamp and found one passible by the assistance of axe men to clear the way along a side hill. In passing along my horses were so much fatigued that they would not drive well and many of them turned down into the swamp from which we extricated the most of them with considerable difficulty. Where I encamped there was no grass for my horses. I was therefore obliged to build a pen for them to keep them from strolling off. Some men sent in the morning for horses returned having found a part of them.

June 5th 1½ Mile North West crossing 2 or 3 small creeks and encamped on a creek 20 yard wide running south west. 2 horses & one mule gave out and were left behind. We had no meat in camp since the morning of the day before and at night I gave out a ration of ½ pint of flour to each man. During the day we hunted hard but saw nothing to kill although there was some Bear & a little fresh Elk sign. At night therefore as we were quite hungry I gave another ration of ½ pint of flower per man and killed a dog the only one we had in camp. For a long time I had been traveling in [our utmost][138a] a country where our utmost exertions would not enable us to travel more than 3 or miles per day at most where my horses were mangled by the craggy rocks of the mountains over which they passed and suffered so much from hunger that I found myself under the necessity of stopping a while to rest them or run the risk of losing many of them if I should proceed.

This situation was verry unpleasant because while my men were suffering from hunger and in a country where there was verry little game they were laying in camp and apparently without the power of supplying their wants the only alternative being patient endurance with a prospect ahead not verry flattering for although near the Ocean yet our intended route appeared equally rough with that over which we had passed. In the vicinity I saw some Beaver sign but the tide setting up interfered with the design of trapping. An affray which happened the day before between one of my men and 2 indians and which I neglected to mention in the proper place was as [follows]:

Two indians following in the rear of our party in company

with one[139] of my men offered him some berries which he took and ate and made signs to them to come on to camp. But they did not understand him and insisted on being paid for the berries he had nothing to give them and they attempted to take some of his clothing by force on which he presented his gun and they ran off he firing as they ran. As he was not a good marksman I presume he did them no hurt. His account of the affair was somewhat different from this but I presume mine is near the truth.

June 6th Remained at the same camp and had some of my men engaged in pressing fur and others hunting. But the hunters after every exertion returned without killing anything. Two of them traveling North West found a pass to the Ocean. Saw some Elk and got a shot at a Bear. As no game could be killed I was obliged to kill a young horse which gave us quite a feast.

June 7th Remaining at the same camp I sent 2 men forward to hunt directing them to encamp where it was my intention to stay the first night after leaving that place. Others of my men were employed in pressing fur and looking for lost horses. 10 or 15 indians visited camp bringing with them a few Muscles & Lamprey Eels and some raspberries of the kind I have before mentioned. In the evening when they left us they stole a small Kettle.[140]

June 8th North West 5 Miles and encamped on the shore of the Ocean at the mouth of a small creek[141] where there was tolerable good grass. The high hills which came in close to the beach of the Ocean presenting a front nearly destitute of timber with bushes breaks and grass in some places. Verry little game for the hunters after the greatest exertion returned at night without having killed or even seen any thing to shoot at. Some of the hunters remained out all night.

June 9th There were several indian Lodges near my camp of whom I purchased a few muscles & small fish. At 11 O Clock the hunters all came in. For three or four days [we][141a] as many of us as considered ourselves good hunters had [all][141b] been employed [ourselves][141c] in hunting. During that time nothing had been killed and but three animals had been shot at 2 Black Bear and one Deer which we wounded. This was what hunters call bad luck and what we felt to be hard times for we were weary and verry hungry.

Among other trifling things which the indians brought us to eat was some dried sea grass mixed with weeds and a few muscles. They were great speculators and never sold their things without dividing them into several small parcels asking more for each than the whole were worth. They also brought some Blubber not bad tasted but dear as gold dust. But all these things served but to agravate our hunger for we were constantly encountering the greatest fatigue and having been long accustomed to living on meat and eating it in no [ordin][141d] moderate quantities nothing else could satisfy our appeties.

In the afternoon I took my horse and rode out to make another effort to kill something to alleviate the sufferings of my faithful party and thanks to the great Benefactor I found a small band of Elk & killed three in a short time which were in good order. I returned to camp and directed several men to go with me with some pack horses without telling them what they had to do. When they came to the spot where the Elk lay their surprise and joy were tumultuous and in a short time their horses were loaded and they returned to camp to change it from the moody silence of hunger to the busy bustle of preparation for cooking and feasting. Little preparation however was necessary when men could be seen in ev'ry part of the camp with meat raw and half roasted in their hands devouring it with the greatest alacrity while from their preparations and remarks you would suppose that nothing less than twenty four hours constant eating would satisfy their appetites.

June 10th Remained at the same camp. My men were employed in making salt and in cutting & drying Meat. Early in the morning the indians came offering to give the Beads I had before traded to them for Meat. I soon made them to understand that what I had to spare which was verry little would be freely given to them.

June 11th North 4 Miles. I packed up early and moved on but missing an ax and drawing knife I stoped the party and searched for them. Not finding them I concluded the indians had stolen them & went back to some indian lodges close at hand. They were all gone but an old man who pretended to know nothing about them. I then went to some Lodges above and when the indians saw us coming they all ran off. But after a while one of them came to me and I told him that I should keep him until the tools were found and at the same time sent the old man found at the other Lodges to tell the indians the reason why their friend was detained.

By searching we found the ax covered in the sand under their fire. The drawing knife could not be found and I took the hostage along tied. After keeping him several hours no indians appearing to relieve him I let him go. The traveling was verry bad and at 4 miles I came to a deep impassible ravine and encamped having to build a pen for my horses lost in the course of the day.

June 12th West 1 Mile. Early in the morning I packed up and moved down a ridge with considerable difficulty to the beach of the ocean where I encamped. I drove the horses across a small creek where there was some grass. The horses lost the day before were found and brought up. 3 of my men quite sick.

June 13th North N West along a ridge in places rough with thickets and rocks. At night descending to and encamping on the shore where there was but little grass. In the course of the day 3 Mules gave out and were left one load was lost and one [mule][141e] horse was disabled by falling down a ledge of rocks.

June 14th North 1 Mile. It being low tide by passing around a point in the water I was enabled to travel along the shore and encamped in a prairae of about 100 acres of tolerable grass. In the vicinity was a plenty of Elk sign. The prospect ahead was somewhat flattering and I was in hope that we had passed all the mountains. Some men sent back for the purpose found the lost load and brought up the fatigued mules.

June 15th I lay by to recruit my horses. Several of us went hunting and Joseph Lapoint in the morning killed one of the largest [animals][141f] Elk I had ever seen. He was not verry fat but [in][141g] tolerable [order][141h] good meat. His size induced me to weigh the meat which I found to weigh 695 lbs neat weight exclusive of the tongue and some other small pieces which would have made it above 700 lbs.

In the evenings hunt Mr. Virgin and myself each killed an Elk not as large as the one before mentioned but one of them was good meat. In the course of the day several indians visited camp bringing some Clams small fish Raspberries strawberries and a Root which on the Columbia is called Commass.[142] These indians traded like those last mentioned.

June 16th North 5 Miles. One mile along the beach north & then turning to the right I traveled 4 Miles across a prairae leaving a range of hills on the East running North not far distant and thickly covered with Hemlock & Cedar. The prairae was covered with brakes bushes & grass & had many springs some of which were miry.

June 17th 1 Mile north at the end of which I came to the termination of the [timber][142a] prairae. Then commenced thick timber and brush and swamps which so much obstructed my progress that I was obliged to retrace my steps and encamp. I then went with one man on the ridge and traveled north 4 or 5 miles when I found it impracticable to move in that direction on account of the thick brush. I returned to camp and sent two men towards the Ocean. When they came in they reported That there would no difficulty in moving to a prairae not more than a Mile distant. Its extent they had not time to ascertain. In their excursion they had killed an Elk dressed & hung it up. Other hunters out killed nothing.

June 18th Remained at the same camp. Sent some men to hunt and others to see which way it would be advisable to travel. The hunters were unsuccessful and [the men] that were looking for a road found it impracticable to travel near the Ocean. They observed a Lake of several Miles in extent along the shore of which it was impassible on account Thick brush and mire.[143]

June 19th Remained at the same camp because I did not wish to move until I knew whether I could find grass for my horses. I took two men and struck East across the ridge following an indian trail 2½ Miles when I struck a river 80 yards wide coming from ESE.[144] I crossed Leaving the Men on the bank and found the river so rapid that my horse fell and it was with great difficulty that I got him out. I went ½ Mile up the river & recrossed having the same difficulty as before. The bottoms along the river were brushy but there were some small prairaes and the hills beyond were bare. On the river was some beaver sign. From the high ground I could easily see that what the men had taken for a Lake was a bay of the Ocean. From the breaking of the water without I supposed its entrance to be shallow.

June 20th East 2½ Miles & encamped on the North bank of the river which I had discovered the day before. It was deep fording. My camp was in a small prairae of good grass. Several men sent hunting.

June 21st North N E 6 Miles. Leaving the river on account of the brush and traveling along a ridge stony and covered with small brush but verry little timber. As I advanced the country became rough and the high ridge on which I was traveling extremely rocky. I saw that it would not answer to move longer and therefore encamped. 3 Deer and a fine Buck Elk killed. Deer verry plenty in the vicinity.

June 22 N North West 5 Miles Being obliged to move again towards the coast I followed a descending ridge for 5 Miles and encamped in a prairae. The country was generally timbered during the days travel but from my camp towards the coast the prospect was generally prairae.

June 23^d North West 8 Miles At 3 Miles I arrived on the shore and from thence I traveled along the shore and sometimes immediately on the beach for 5 miles and encamped after crossing a creek 20 yards wide. The hills came within ½ mile or a mile of the sea, and were generally bare of timber. The Low land along the shore and in the valleys covered with high breaks and has some Miry springs. Many indians visited camp in the evening bringing berries small fish and Roots for trade. [During the d][144a] In the course of the day one Mule gave out and another ran back on the trail.

June 24th West North West 3 miles and encamped at the mouth of a river 50 yards wide rapid at the mouth but as it was high tide I could not cross.[145] The hills about the same distance from the coast as the day before and the low land thick covered with brakes scotch caps and grass. When starting in the morning I sent two men back for the Mules that had been left the day before. They came in the evening without the mules and I immediately sent two men back but they soon returned as the indians at a village close at hand did not appear friendly.

Near my camp was a village of 10 or 12 Lodges but the indians had all ran off. Among the indians of this country I have seen a small kind of Tobacco which is pretty generally cultivated. These indians Catch Elk in Pits dug in places much frequented. They are 10 or 12 feet deep and much Larger at the [top][145a] bottom than [bottom][145b] top. They are completely covered over and some of my hunters with their horses fell into one and got out with considerable difficulty.

June 25th North N West Early in the morning as it was low tide I packed up and forded the river. During the principal part of the days travel the hill came in close to the rocky shore. I was therefore obliged to turn out into the hills which were nearly bare of timber but brushy and cut by some dark ravines. In the morning when my horses were brought up I found two of them wounded with arrows and in the evening one was missing which I supposed to have been killed. This day I traveled 12 Miles which was much the best march I had made for a long time. Deer plenty and some Elk.

June 26th N N West 8 miles. On leaving camp I struck out from the Ocean following a ridge on a circuitous route [for][145c] until It came into the Ocean again at the mouth of a creek 20 yds wide where I encamped. The place from which I started [was][145d] in the morning was covered with brakes & brush when I got out among the hills I found some timber & good grass & where I struck the [shore][145e] a sandy soil short grass low Pines Sand Cherries and strawberries.

June 27th North 7 Miles. With the exception of two or three steep points which I was obliged to pass over I was able during the day to travel along the beach. I encamped on the south side of a bay and close to its entrance which was 150 yards wide. The Bay itself was 3 Miles long and 1 Mile wide. At low water I found it quite fresh, from which circumstance I infered that it received a considerable river.[146] After encamping I made rafts that I might be ready to cross the bay early on the following morning. On each side of the Bay were several indian villages but the indians had all run off. On a creek which I crossed 3 miles back was some beaver sign and also some in the bay.

June 28th N N West 6 Miles. Early in the morning as it was low water I commenced crossing. And when I had finished

I had lost 12 or 15 drowned in the middle of the water. I know not the reason of their drowning unless it might perhaps be ascribed to driving them to much in a body. In three days I had lost by various accidents 23 horses & mules.

June 29th N N West 5 Miles. The traveling for the last two days much alike alternately on the beach and over the hills which generally closed in to the shore near which the country was generally prairae with some thickets. Farther back from the coast the hills were high rough and covered with thickets & timber. This day I could have traveled farther had it not been high tide which prevented me from traveling on the beach and the hills were too rough to allow me to leave the shore. In the vicinity of my camp the country was clothed with fine grass and other herbage, a good grazing country though somewhat rough.

June 30th North 5 Miles. After traveling 2 Miles I was obliged to leave the coast and travel over the hills to my encampment which was a short distance from the shore where there was good grass. From a high hill I had an opportunity to view the country which Eastward was high rough hills and mountains generally timbered & north along the coast apparently Low with some prairae. In climing a precipice on leaving the shore one of my pack Mules fell off and was killed.

July 1st 1828 North 9 Miles. At 5 Miles from camp crossed a creek the outlet of a small Lake[147] on which was some Beaver sign. At this place the hills recede from the shore leaving a bluff from 30 to 100 feet in heighth. Immediately on this bank is a narrow skirt of prairae and further back low Pine & brush. The soil thin and loose. Encamped on a river 60 yards wide on which was some beaver sign.[148] I found the tide too high to cross. For the three past days but one deer had been killed but as we had dried meat we did not suffer from hunger. We saw appearances of Elk have been abundant in the vicinity when the grass was tender. For many days we had hardly got sight of an indian and but one had visited camp since my horses were killed. In the course of the days travel one of my horses was crowded off from a cliff and killed.

July 2nd 12 Miles North principally alond the shore at 6 Miles from camp passing a small Lake.[149] During the days travel the hills were generally 3 or 4 Miles from the shore the intermediate space being interspersed with grassy prairae brush, sand hills & low Pines.

July 3^{d} 5 Miles N N West. At 2 Miles from camp I came to a river 200 yards wide which although the tide was low was deep and apparently a considerable River.[150] On first arriving in sight I discovered [two] some indians moving as fast as possible up the river in a canoe. I ran my horse to get above them in order to stop them. When I got opposite to them & they discovered they could not make their escape they put ashore and drawing their canoe up the bank they fell to work with all their might to split it in pieces.

[The *Journal* ends with the entry of July 3rd][151]

The front cover of the Jedediah Smith transcript Journal

NOW the Smith party was augmented by addition of Marion, an Indian boy from the Willamette Valley who had been a slave in the band frightened by Smith on July 3, and who had been left behind when the rest ran away.

Word traveled from village to village ahead of the expedition that a company of hostile white men was passing through the country. Horses were shot furtively with arrows, but the natives were so numerous no reprisal was attempted. On July 10, Smith observed the Indians he met were acting as though they planned to attack him, and, accordingly, vigilance was increased.

Two days later the party crossed the Umpqua River below the mouth of Smith River, in what is now Douglas County. Half a hundred of the Kelawatset tribe gathered at the camp of the travelers. They had in their possession British trade goods, and their familiarity with the names of the British leaders made the American party feel more secure. Besides, a Hudson's Bay Company post, Fort Vancouver, was relatively not far away.

One of the Kelawatsets stole an axe. He was seized and tied with a rope while the Smith men stood with rifles ready in case of an attack.

Frightened, the prisoner confessed where this valuable tool lay buried in the sand. He was, however, one of the principal men of the tribe and upon his release endeavored to stir his fellows to avenge his humiliation. Upon the counsel of a chief of higher rank his plan was rejected.

Then, according to the Indian side of the story, this chief of the conciliatory disposition himself was insulted. He climbed on one of the expedition horses for a ride around the camp; but Arthur Black compelled him to dismount. Thereupon the peacemaker agreed to the proposed assault.

The American party was on the west bank of Smith River. Captain Smith, with John Turner, Richard Leland and an Indian guide, went up-stream in a canoe, looking for a suitable place to cross the horses. Before he went the captain warned his men not to let the natives come into camp.

Harrison Rogers evidently was deceived by the Kelawatsets' apparent friendship with the whites of the Hudson's Bay Company. Without the knowledge of his captain he allowed the Indians to mingle with the men in camp. The result was fatal to him and all the rest save Arthur Black.

The Indians, 200 strong, fell upon the travelers. Black ran to the woods and escaped. After much suffering he was taken to Fort Vancouver, by the friendly Killamoux, or Tillamook Indians.

While the slaughter of his men was taking place, Smith was on his way back to camp. Suddenly the canoe was overturned by the Indian guide. Jedediah and his two companions managed to swim ashore under fire from the opposite bank, and make their way northwest to the Tillamook River and thence to the fort, where they arrived on August 10, 1828.

Dr. John McLoughlin,[152] chief factor at Fort Vancouver,

sent an expedition to recover the American property. Alexander R. McLeod, a chief trader, was in command. Jedediah Smith and his remaining men joined the expedition.

McLeod went from the Columbia River down the valley of the Willamette to the Umpqua, thence to Smith River—named Defeat River by Smith—thence to the seacoast and along it both north and south a short distance, collecting Smith's goods and horses, which had been scattered among various villages.

Fragments of the journals of Jedediah Smith and Harrison Rogers were recovered. Details of this expedition are given by McLeod in his own journal, which follows, after two introductory letters.

* * *

Letter of Chief Factor John McLoughlin to Jedediah S. Smith, Fort Vancouver, 12th September, 1828.[153]

Fort Vancouver,
12th Sep$^{r.}$ 1828

J: S: Smith Esqr.,
Dear Sir,

I am extremely sorry to learn from Michel [La Framboise] that your property is so scattered that their is little probability of recovering it, you write "In the meantime should you think it necessary for the benefit of your Company to punish these Indians you would confer a favour on your humble Servant to allow him and his Men to assist" I beg to assure you that in this case I am actuated by no selfish motives of Interest—but solely by feelings of humanity as I conceive in our intercourse with such barbarians we ought always to keep in view the future consequences likely to result from our conduct as unless those Murderers of your people & Robbers of your property are made to return their plunder, as we unfortunately too well know they have no horror or compunction of Conscience at depriving their fellow Man of Life—If strangers came in their way they would not hesitate to murder them for the sake of possessing themselves of their property, but as it would be worse than useless to attempt more than our forces would enable us to accomplish and as Mr. McLeod knows those Indians & knows best whether we can effect any good, he will decide on what is to be done most sincerely Wishing you succees, Believe me to be

Yours truly
Jno. McLoughlin
Chief Factor
Hudsons Bay Co^y.

* * *

Letter of Chief Factor John McLoughlin to Chief Trader Alexander Roderick McLeod, dated Fort Vancouver, 12th September, 1828.[154]

Fort Vancouver,
12th Sept. 1828

A. R. McLeod Esqr.,
Dear Sir,

I received yours of the 8th per Michel Laframboise & am extremely sorry to find by his Statement that Mr. Smith's affair has a more gloomy appearance than I expected & it seems to be in that state, either that we must make War on the Murders of his people to make them restore his property or drop the business entirely.

I know many people will argue that we have no right to make war on the Natives, on the other hand if the business is drop[p]ed, will not our personal security be endangered wherever this report reaches—Again suppose that by accident a Vessel was wrecked on the Coast, to possess themselves of the property would not the Natives—seeing these Murderers escape with impunity—kill all the Crew that fell in their power & say as these now do—We did not take them to be the same people as you—have not the Natives of Cape Look-out not many years since killed the Crew of a Vessel wrecked opposite their Village, and is it not our duty as Christians to endeavour as much as possible to prevent the perpetration of such atrocious crimes—& is their any measure so likely to accomplish so effectually this object as to make these Murderers restore at least their illgotten booty now in their possession—But it is unnecessary after the various conversations we have had for me to say any thing further on this subject—You know those Indians you know our means, and as a failure in undertaking too much, would make this unfortunate affair worse—& as you are on the spot—you therefore will decide on what is best to be done and depend that whatever that decision may be at least as far as I am concerned every allowance will be made for the situation you are placed in.

I am Sir

Yours truly

Jno. McLoughlin.

N. B. Laframboise and [Joseph] Cournoyer will go to the Umqua or return immediately as you think proper—Mr. Smith offers himself and party to accompany you to War on the Murderers—I refer him to you for an Answer.

Front cover of Journal of Alexander R. McLeod

Alex R. McLeod's Journal
Southern Expedition

Particular Occurrences during a Voyage of about three Months, Southward of the Columbia[155]—

SEPTEMBER—SATURDAY 6th Fine Weather —At 4 P. M. left Fort Vancouver, in a Boat with Six Men having a Canoe in Company, both much incumbered with Baggage, the Boat so indifferent as to require a Man Constantly Employed baling out Water—shaped our course towards the Wullamette River, and fixed our Camp for the Night on the 1st Point of its entrance on the East shore—Mr. J. Smith accompanies us—

SUNDAY 7th Fine Weather—with the days dawn we were in Motion, being then about five O'Clock—at 4 P. M. reached the *Chuttes,*[156] having come a distance of about five and twenty miles—i. e. four miles from the Entrance of the River to the junction of both Channels—seventeen to the first Rapid and four to the Portage of three hundred yards over which our heavy baggage was carried by dusk the Indians assisted and each [received] a bit of tobacco, it was considered ample remuneration —found a Boat which was left for us—the Men having been oblidged to return for want of Provisions—few Indians at this place the majority being up the Country. various reports of our horses, these people like the Generality in this Quarter are so subject to exaggeration that their assertions are entitled to little credibility therefore their stories treated with indifference.—

MONDAY 8th Fine Weather. about six A. M. all our Baggage put on board of the Boat, being oblidged to leave the Canoe. we proceeded and reached *Sampou yea*[157] at 2 in the afternoon about a Mile short of the place, met Mr [Thomas] M.[c]Kay,[158] [Michel] La Framboise &

three Men in a Canoe, who returned with us, the latter arrived to day from the Umpqua—he had no personal Communication with any of those Indians, he returned from near the Old Fort,[159] Nasti who accompanied him, acted as Linguist carried a Message to the Principal Chief of that Tribe who seems still to value our support and seems exasperated at the late action of the *Keliwat-set* Indians who defeated Mr Smiths Party—some of the horses have been taken by the Umpqua Indians, who still possess them, all the Property furs as well as other things is dispersed over the Country—La Framboise on his return met with some of those Skins in the possession of a Wullamet Indian and could only recover them in the ordinary way of trade, from an other Individual of the same tribe he got one of the horses in the same way—all of which we delivered to Mr Smith—Distance and courses as follows, 1½ Mile S.S.W. a Small River runing in from the West ½ a Mile above the Falls, S.S.E. 1 Mile. S. 1¼ Mile. S.E. ¼ Mile. S. 1 2-3 Miles. West1 Mile. N.W. ½ Mile. W. ½ Mile. S.W. 1 Mile. N.N.W. ½ Mile. W. ¼ Mile. W.S.W. ½ Mile. S.W. 1½ Mile.—W.N.W. ½ Mile. S.W. 2½ Miles. S.S.W. ¼Mile. S. 2½ Miles. S.S.W. 2½ Miles. S.W. 1 Mile. W.S.W. 1 Mile. S.S.W. ¼ Mile. this days route from the Falls to the Site of the Old Establishment where our peoples Camp is situated—horses much reduced principally thru [having been] left here since last March, and no proper person to Guard them from the Molestation of the Indians—of late the fire has committed such ravages that Scarcely any feeding is left for our Animals—Gave the Men their Regale.—[160]

Tuesday 9th Fine Weather. La Framboise & Party proceeded to Fort Vancouver, to make his report to C.[hief]F.[actor] McLoughlin, I expect his return in four days and intimated the Same to my Senior Officer—the Men, such as were not too much affected with liquor, employed making Saddles—

Wednesday 10th Fine Weather.—the Men employed as yesterday, many unable to do any job, from the effects of Liquor all the horses that could be found were brought to Water some Still missing every endeavour to find them failed *Caisino* and Suite proceeded to the Southward, on a visit—

Thursday 11th Fine Weather. Preparations to Start going forward, and nearly complete, horses attended to as usual. —the Men Still enjoying themselves, but I am glad to observe that little Liquor now remains in their possession.

Friday 12th In the forenoon Cloudy Weather and Rain succeeded in heavy showers—this circumstance effectually stoped our progress. Mr. Smith shot a Small Deer, which happened very oportunely as we had no Venison—an attempt made to assemble our horses failed. we anticipated La framboise's arrival, but were disappointed.—all our Baggage is ready—traps are given to the Men with every other necessary Article for hunting—

Saturday 13th the Weather continued over cast during the night, and cleared up after day light and continued fine the remainder of the day—Horses collected and distributed to each man in proportion to Quality, this occupied us the most part of the day and we had to postpone our departure till tomorrow.—had the Weather permitted, we would have started a Party as feeding for our Animals is very Scanty—at six P.M. M. La framboise & Party arrived from the Fort [handed me a letter from C. F. McLoughlin intimating a desire to use every means to warrant the restitution of Mr Smiths Property—leaving it at my option to take these two

Men with Me. I availed myself thereof as our Number can't be too great in the eyes of the Natives—],[161] No news of Consequence.—Several Shots fired at Deer, but none Killed—

SUNDAY 14th Same Weather as yesterday, of course we could not move—further exertion made to recover our Strayed horses, but without Effect—towards evening the Weather bore an indication for the better—

MONDAY 15th Light Rain at intervals, all hands turned out early in the Morning to Seek our horses and after Noon I proceeded with a Party, leaving others [including two of Smith's men] with Mr. McKay, to endeavor and find the Strayed horses our route led Southward, the Want of Grass, made us go till 9 P.M. when we reached a Small River where there is a little Picking for our Animals. Many horses strayed after dusk, some with their loads —it past 11 at Night when the rear came up

TUESDAY 16th Rainy Weather—Sent back Men and horses, to where Mr. McKay is as part of our baggage remained there—other Men hunted for the horses that Strayed from us last Night and recovered Many yet a number are Missing principally *Marrons*[162]—Hunters out but no Success—

WEDNESDAY 17th Cloudy Weather—In the Afternoon Mr. McKay & his Party arrived, having found the Strayed horses, with the exception of four, that were not Seen Since my arrival and We suppose them Stolen by Indians inhabiting the Vicinity of Mount Hood—*Caisino* arrived from *Sandiam* River[163] and now proceeds to the Columbia—hunters out much Shooting two Deers, Killed and some Ducks—three young horses have remained in our rear and will be sent after in course of tomorrow—

THURSDAY 18th Rainy Weather, so we could not rise Camp.

Caisino departed on his return to the Columbia, by him addressed a few lines to Chief Factor McLoughlin—a Party of Men made a fruitless attempt to recover four Marrons Strayed on our way from *Sampouyia*—a few Deer Killed—

SEPTEMBER, FRIDAY, 19th The rain continued at long intervals all day, tho our horses were collected to make a Move, we were obliged to turn them to the field—5 Deer Killed, some Indians came about us—but had no furs—Louis Shanagorate came from River Sandiam, having past the Night with us returned to his Camp this afternoon—

SATURDAY 20th less Rain than yesterday, but our horses having Strayed in the Night, we lost the day to collect them—hunters out, no success—

SUNDAY 21st Flying Showers of Rain, Rised Camp Course South Distance Twelve Miles. past River *Chembukte,*[164] In the Afternoon [Joseph] Gervais, [Jean Baptiste] D'Epatis, Louis & Jacques, Met us at our Camp, they bring us unfavourable accounts of the Umpqua Indians, it is the common report of other tribes, that the former are ill disposed towards us, having received large presents from the *Kelewasets,* who defeated Mr. Smiths People, are now resolved to support them, elated with their late success, they expect to make an easy capture of us, as soon as we enter their Country, it is said they are mustering Strong to way lay us in the Woody parts of the Country, Pillage is their object unmindful of the Consequences, this is the substance of the report brought us by the free Men, and they got it from the *Lamali* Indians who have visitted the Umpquas. Encamped at the little River, near half ways over the Point of high land

MONDAY 22nd Fine Weather, Rised Camp Distance about 8

Southern Expedition

List of Freemen Engagés and Assistants

1	Alexis Aubichon	Nasti
2	Canasanarette	Soloquin
3	J. Bt. Dubreuille	Antoine
4	Joseph Gervais	Blondin
5	J. Bapt. Jn. Louis	Chumbeck
6	J. Bapt. Depatie	
7	Chas. Jeaudoin	
8	Willm. Johnson	
9	Pierre Mokonnquian	
10	Pierre L'Etang	
11	Michel Otodanie	
12	J. Bt. Perrault	
13	Amable Quesnelle	
14	H. Le Dorante	
15	Louis Shoegankatse	
16	Charles Tehgte	
17	Ignace Shatectoanie	
18	Louis Shorregarate	
19	Thomas Tawatom	
20	Pierre Villandree	
21	Michel La Framboise	
22	Jos. Cournoyer	
	Cartonache	
	Mon Cousin	
	Jack	
	Jacques	
	John	
	Charles	
	Callepouzia	
	Baptiste	
	Verrtau	

Alexander R. McLeod's record of men under his command, on the expedition to recover Smith's property.

Miles, to River Sandiam, where we encamped, the Water being Still high for our horses to ford the Channel, loaded, and the Water falling fast, induced us to wait till to morrow—Frisé & little M'ichel [Otodanic?] are unwell the latter having a Boil, on the inner side of the Thigh so that he can't ride, yet he must walk, with great Pain and difficulty—the former Complains of a Pain in the Breast which makes riding insuportable—they both reach[d] camp late— [Charles] Jeaudoin has very sore Eyes—some Indians brought Skins to trade, which we rejected, referring them to Laframboises Return—here we found Louis and Jacques Camp—much trouble with Marrons..

TUESDAY 23rd Fine Weather—Continued our route, Course the Same distance 18 Miles—Encamped at River Coupé[165] on the North Bank—several Deer Seen, none Killed—Our Wild horses cause a great deal of trouble, in fact I begin to think we will have to abandon them, as they cause much trouble and delay—Indians came to us with roots to trade—

WEDNESDAY 24th Fine Weather—Continued our progress, about 12 Miles. Course S. & W. Encamped on the West Shore of the Wullamette, fording the River about Knee deep Water, the Marrons caused three loaded horses to rush into the Deep, two were drowned loaded with traps & lead—the third loaded with Grain got safe across, every endeavour was made to recover the lost Articles without Success—Night put a stop to further exertion. the Indians seem shy of us—

THURSDAY 25th Fine Weather, Sent People to get Canoes and conveyed Word to the Indians, in the vicinity, to try their Dexterity in Diving, a Blanket and an ax, was Stipulated as a remuneration to the Individual who would succeed in finding the horses or Property—In the Afternoon, one of the Indians, luckily found One, with 19 traps out of twenty, One supposed to have remained at the Bottom—the other horse is supposed to have been seen, some Individuals seem to be possitive of having seen it, so that bright hopes exists of recovering our Property in course of tomorrow—this Evening many Shots fired at Deer, One Only Killed—

FRIDAY 26th Fine Weather, Men and Indians employed as yesterday, but with equal Success as the attempt proved fruitless, and the close of day put a Stop to further exertions, the Water is very Cold A Circumstance which ad[d]ed to the Rapidity of the Current and the Many trees fast at the Bottom, is much against the Divers, a Kind of Grapple was tryed without producing the desired Effect, the Depth of Water may be about twenty feet, but in a particular place it exceeds that, Understanding that Gervais had a Slave that was famed for Diving I sent for him, he cast[?] up about 9 P.M.—the Stipulated reward was given to the Indian, that found the horse, this created further exertions, and the hope of Gain held out as an encouragement to persivere—One Deer Killed—

SATURDAY 27th Fine Weather, throughout the day persivered in our Search, but ineffectually—the Indians again joined from the exertion Made justify's the Opinion that the Animal drifted with the force of the Current —Mr. S.[mith] who past the Night from Camp returned with a Deer—

SUNDAY 28th Fine Weather. the Indians wishing to have their Canoes, to enable them to go in Quest of their food, left us without Means of Continuing our Search, consequently the Men were divided into three Parties, and directed to make three Canoes, at which they were employed all day—

MONDAY 29th Fine Weather—at Noon two Canoes being ready, reassumed the Sea[r]ch till Dusk, without any Success. In the Afternoon a Party of Men, Started for the purpose of hunting.—no Animals in our vicinity, they must proceed to the hills, and in consequence of the distance have to sleep from Camp.—yesterday Evening D'Epatis came to our Camp, and assisted the Men, in their endeavours to recover our lost property unsuccessfully—

TUESDAY 30th Cloudy Weather—About Noon M- La Framboise & Party arrived, their delay was occasioned by the Indians being dispersed in detached Parties in various directions, remote from each other, and as the object of the Party was to obtain horses, as many as possible, much time was lost to visit the different Parties of Indians, and 8 horses only were traded rather at a more extravagent Price, than we expected, and only our present Situation could make us pay them so high, Still we require more to make us independent, as once out of this Section of the Country, no means exists of procuring any—the Men that Slept from Camp returned with Six Deers—those that remained at Camp, Continued Searching for our lost property, but in Vain—

OCTOBER, WEDNESDAY 1st Fine Weather—Every endeavour to find our lost property proving fruitless, gave us all hopes, and rised Camp along River *Nomtom ba,*[166] distance 15 Miles, Course S. & S.W. Saw Several Deer, some Shots fired without Effect.—forded the River and Encamped on the East Shore—In the Evening D'Epatis left us to join his family he and Gervais are directed to meet tomorrow with their followers, at an appointed place.—

THURSDAY 2nd Fine Weather, Continued our route, Course South Distance 9 Miles, along the Banks of the Same

River—D'Epatis & Gervais with their followers, are now attached to our Party, forming in all twenty Men, nearly as many Slaves, besides Mr. S. & his three Men, which in the Eyes of the Natives, makes a forcible impression if we can judge from the alarm a few Indians s[e]en in course of the day, got. they instantly Sheltered themselves in the woods intrities of their Acquaintances appeased them and they came to us afterwards quit[e] composed.—hunters out but no success—

FRIDAY 3rd Fine Weather, Started at the usual hour, when our horses got fagged put up. Distance 15 Miles Direction South—Entered the Mountain, bad Water and no Grass—hunters traveled much no Success—

SATURDAY 4th Fine Weather, La Framboise with two Men, proceeded to the Eastward, to recover a horse belonging to Mr Smith, Said to be in possession of Indians in that Quarter Proceeded Six Miles and encamped finding plenty of Grass and Water for our Horses, many of whom are very poor already—No Deer Seen—no Indians seen in this Section of the Country, supposed to be occasioned by a general alarm among the different tribes ,and we know not where they have fled to.—

SUNDAY 5 Fine Weather, Continued our progress in a Southerly direction till over the Mountain La Biche then Westerly along the River of the Same Name,[167] forded it and pitched our Camp on a fine Plain, where my Camp stood two months last Winter—Distance today 18 Miles, according to Indian report we were led to expect Seeing Indians in this Vicinity, it was even asserted and believed by some of our Party, that, they would way lay us, in Woods through which we past in course of to day but that like many other Indian stories proved Groundless as we have not discovered even late Vestiges —three Deer Killed the first we got since leaving the

Wullamette River—Gervais tried the Woods for Elk and did not come to Camp—

MONDAY 6th Fine Weather—All hands who have any pretention to hunting, tried their Skill, many shots were fired and only five Deer Killed—Gervais returned this forenoon, he say [*sic*] two tracks of Buck Elk, but could not over take them. La framboise arrived and brought the horse he went for, the Property of M^r^. Smith accordingly made over to him.—Indians La framboise went to, informed him that those of the Umpqua, hearing of the Strength of our Party, and supposing our intention hostile, got intimidated and were all of[f] to the Mountains, this is a new version and likely without foundation. *Charles* and *Toloqua* will be here tomorrow, their Errant is for traps formerly lent them and now Claimed in case of need.—No Indians discovered as yet—La Framboise had to leave his horse on the Way.

TUESDAY 7 Fine Weather.—Toloquois & Charlos arrived brought traps—Indian vestiges of a late date Seen not far from Camp, on the Mountains, supposed by Spies watching our movements—a few Deer Killed.

WEDNESDAY 8th Fine Weather, In consequence of late information Stating the Umpqua Indians to be collected, with the Chief at their head, and Stationed at the Site of the Old Establishment on Said River, and those people possessing horses belonging to M^r^. Smith and our route being in that direction, we proceeded forward, Course as usual Southerly Distance 18 Miles, encamped on a Branch of the Umpqua, coming from the N.E.—in course of the day saw Some Indians a head who took to the woods as soon as they saw us, as it was of importance to our object to have a Communication with them an Indian of our party was sent to them with a suitable Message which had the desired effect and confirmed the former Story relative to the Indians being assembled at the Same place as above Stated—saw other Indians on the summit of the hills none came to us till after we were encamped their assertions tends to confirm what others told us with the exception of the residence of the Chief and Party who are Some distance to the Eastward of the site of the Old Establishment. they have some horses but the number our informants can not tell, Many have been killed by the Natives on the Coast and a loss sustained in conveying them from the Sea—a Bear and a Couple of Deer killed.—Bears said to be numerous of the Grizzly Kind.—it is worthy of remark that most of the Men composing the Party are ailing in some Shape or other some lame—others Sick some with sore Eyes not five Sound in Health—

THURSDAY 9th Fine Weather—Continued our route Distance 11 Miles, Course South, Encamped on the Umpqua River, on the North Side near the Site of the Old Establishment Some Indians residing in the Vicinity fled at our approach, but were soon made to return, as their Countrymen accompanying us appeased their fear—and they returned to their habitations treated them in the customary way—a few Deer Killed.

FRIDAY 10 Fine Weather—Sent a Message to the Chief to come to us and is expected in a day or two.—Indian intelligence purporting that four of Mr Smiths Men are in the custody of *Cahoose*[168] Indians, how these people have escaped, we are left to conjecture, Several Indians affirm that they are in existence, if the Old Chief confirms the report [it] will remove my doubts on the Subject. two horses and a mule, were observed on the Opposite Shore and were brought to Camp being Mr. S's Property—one of our lads States to have Seen an other while hunting on the South Shore, a few Deer Killed.

SATURDAY 11th Fine Weather—the Umpqua Chief with a Dozen of his tribe arrived, they have brought 8 horses restored them to their owner M[r]. Smith—had a conversation with this Leader, St. Arnoose, who has been in person on a visit to the Kellywasats after they defeated M[r]. Smiths Party, and we enquired into the Cause that gave rise to that unfortunate affair, and the Old man Stated, that while M[r]. Smiths people were busy fixing Canoes together by means of Sticks, to convey their Baggage over the Channel, an ax was mis[s]ed and suspicion led to suspect the Indians of having embe[z]zled it consequently to recover the Property an Indian of that tribe was seized tyed and otherwise ill treated, and only liberated after the Ax was found in the Sand, this Indian happened to be of Rank, of course much irritated at the treatment he met with, declared his intentions to his tribe, to retaliate on the offenders, but he was overruled, by an other Individual higher in Rank and possessing greater influence, subsequently this same man wishing to ride a horse for amusement about the Camp took the liberty of mounting one for the purpose when one of M[r] Smiths men, having a Gun in his hand and an irritated aspect desired the Indian angrily to dismount, the Indian instantly obeyed, hurt at the Idea and suspecting the Man disposed to take his life he gave his concurrence to the Plain in agitation in which dicission, the Indians were much influenced by the Assertions of the other Party, telling them that they were a different people from us, and would soon monopolize the trade, and turn us out of the Country these Circumstances and harsh treatment combined caused their untimely fate, at a moment the[y] least expected it.—the property the Indians got is all disposed of along the Coast—Our Informant can't say anything possitive regarding the 4 Whites Said to be in the neighbourhood of River *Shiquits* or Cahoose, he having heard of it merely as a flying report, from that Quarter through the Interior—we requested him to endeavour and obtain what information he could Glean on the subject.—understanding, that Several horses were Still to be recovered in this Section of the Country, we defer[r]ed remunating the Indians till we had got all—intimated our wish to the Chief to interest himself therein, which he promised to do, and proposed to start on the Morrow on that Mission—2 Bears & 9 Deer Killed—

SUNDAY 12th Fine Weather.—Four Men went off on a hunting excursion wil return tomorrow—St Arnoose departed with one of our Young Men, agreeable to his promise of yesterday, various reports propagated by the Indians about here, relative to the Disposition and intention of the *Kellywasat* tribe, the Old Chief enquired if we intended to make war which we answered in the Negative, knowing his disposition towards them of old, he and his Nation would readily take arms against them more especially if supported by us to enhance his own merrit—being told our wish was to Establish Peace and Quietness and recover what could be got of M[r] Smiths Property, and restore the Same, seemed to give Satisfaction, but privately with some of his intimate Acquaintances the Old fellow, entered more minutely into the subject, and expressed his surprise at our interference in aiding and assisting People that evinced evil intentions towards us, as he had been informed by the people who defeated the Party, they having communicated something about territorial Claim, and that they would soon possess themselves of the Country, makes the Natives about us very inquisitive not having ever heard such a thing before, and we avoid giving them any information, and treat the subject with derision. M[r]. Smith when told of this, observed that he did not doubt of it, but it was without his knowledge and must have been intimated to the Indians through the Medium of a Slave boy attached to his Party, a Native of the Wullamette—he could converse freely with those Indians—as to the Origin of the Quarrel as Stated yesterday by the Old Chief M[r]. Smith affirms to have tied an Indian and set him free when the ax was restored, but denies having used blows or any manner of violence except Seizing him—[Arthur] Black acknowledges to have seen a Chief mount a horse without leave and ordered him to desist but not in an angry tone neither did he present his Gun, but had it in his hand, and he adds the Indian immediately dismounted, shortly after the party was attacked and defeated—

Monday 13th Fine Weather—hunters returned with the meat of a Gra[y] Bear & two Deer two other Deer were Killed close by in the hills, the Old Chief returned with ten horses—M[r]. Smith has received up to this date 26 horses and Mules—one of the Number the Old Man brought today, was left on the way, from excessive weakness—an other is at too great a distance to be got at now, however, they will soon be recovered—

October Tuesday 14th Fine Weather. three Men got leave to Sleep out for the purpose of hunting will return tomorrow—the Old Chief got Several Articles as a remuneration for his Services, which he distributed among his followers, who assisted in bringing horses up from the Coast, they appeared satisfyed and the Old Man unasked proposed to accompany us on our intended journey to the Sea Coast—and at our request promised to be here in two days with Six Canoes to convey the Party from the *Verveau,*[169] to the Sea, this precautionary Measure is taken in case of not finding Canoes on the route, the hire of each costs us a half Ax—in course of last Night 4 Natives attached to the Party, in the Service of the free Men deserted, three of them having Guns were wise enough not to leave [them?], besides helping themselves to Some Ammunition and other Small Articles, their intention is to return to the Wullamette—at present we have no leisure to Send in quest of them, as it would take much time, and likely to no purpose—

WEDNESDAY 15th High Southerly Wind—indicative of a change of Weather.—the hunters who Slept out returned brought three Deer and a Bear— others who hunted from Camp brought in four more Deer to Camp—a Messenger was dispatched with a letter to those people reported to be in the Neighborhood of *Cahoose* or Shiquits River, M^r^. Smith is of Opinion that [Ephraim] Logan & three other Men [James Scott, Jacob O'Hara and William Bell][170] have not visitted their Deposit last Season, and were proceeding towards his route, and possibly fell on his track and and have come forward till their progress was arrested by the Natives, the probability of this Conjecture having some foundation, an Indian acquainted with the route through the Interior, was hired to convey them information of our endeavours to afford them every assistance in our power, that our means will warrant—other Indians as well as the One we engaged to have heard the Same Account of those People as above narrated—

OCTOBER THURSDAY 16th Some Rain fell in the Night, but the Weather Continued fine all day—The Old Chief agreeable to promise arrived brought the Six Canoes as Stipulated—Five of M[r]. Smiths horses got in to day and delivered to him—Preparations making to Start on the Morrow for the Coast M[r]. Smith with a Man is to be of the Party making a total of Twenty—the others have enough to do about Camp under the Charge of M[r]. T. McKay.

17 Left the Camp in charge of M[r]. M.Kay and a Party of Men and Started with La framboise and Sixteen Men, accompanied by M[r]. J.S. Smith & one of his Men—The two others left with M[r]. M Kay to take care of their horses—our Canoes Navigated by Indians, the Old Chief volunteered to be of the party and took charge of the Crafts, we took horses to facilitate our return as by Water it would prove tedious and delaytory—Our route led over a Neck of Land, very hilly, Course Northerly, distance 15 Miles Encamped on the Main Umpqua River—

18 Continued our progress 12 Miles and Encamped, Course west of north, our track led along the River in many places Country hilly and occasionally thick woods.

October 19th Raised Camp Distance 8 Miles—Course west of North Some Deer Killed. Saw the Carcasses of Several horses, Killed by the Indians a long time ago—the Chief and his followers came up with us about Dark, who confirmed our Conjectures about the horses, he was of the Party.

20th Hitherto we have enjoyed fine Weather, but a Change has come on Suddenly and heavy Rain Continued during the day. Some Deer Killed gave a liberal Share to the Indians accompanying us. Raised Camp past River La Biche at its junction with River Umpqua, Encamped below *Grand Cote* and availed ourselves of the leaves of trees for a Skreen against the Wet—Distance 15 Miles—Course Westerly

21 Heavy Rain Continued—proceeded about 12 termination of the Open Country, within about 1½ Mile of the Village pretty populous, on observing a Couple of Graves newly erected excited our Curiosity and on enquiry of the Indians in Company, they told us it was two Individuals of the *Ds-alel* Indians Killed in the fray by the Party defeated by them—In the evening Sent a Message by the Chief, *Starnoose,* to the Village requesting restitution of the Property in their possession belonging to Mr Smith—

22 Constant Rain till in the afternoon it Moderated—Indians from the Village restored, 1 Rifle, 2 Pistol, 1 Musket, Some Books and other Paper, Charts, 2 Vials Medicines, 139 Large Beavers, 24 Small—22 Large Land Otters, 20 Small, 1 Com. Cotton Shirt—1 Russian Ditto —½ Doz led Pencils—

23 Weather fine—The following Articles restored viz 421 Large Beavers—4 Small—1 Large land Otter 4 Sea Otters.—

24 Endeavoured to get Canoes and Indians to Convey the we have on hand to our Camp, but could not get any Sizeable Craft—

Octr 25 & 26 Occupied in getting the furs &c lately recovered under way to our Camp under the Charge of ours and one of Mr Smiths Men—

27 With the remainder of our party forming a total of 16 we proceeded down Stream in three Canoes, took a position opposite to the Second Village—these people immediately restored what they possessed of Mr Smiths Property to wit, 10^lb^ Beads—1 Steel trap 1 Fowling piece, 1 Musket & a Cooking Kettle—

28 Fine Weather—Proceeded to the Sea—Stoped at the entrance of the North Branch, where Mr Smiths Party were destroyed, and a Sad Spectacle of Indian barbarity

presented itself to our View, the Skeletons of eleven[171] of those Miserabl Sufferers lying bleaching in the Sun, after paying the last Service to their remains we continued forward and made the Coast, no Indians in the Vicinity, contrary to their former Custom as several Villages used to be about this place, the Natives are now more Collected than formerly—

29 Fine Weather—Secured our Crafts and such Effects as we could not carry with us—and proceeded along the Coast to the Nor[—?]—about Six Miles off we came to a Small River,[172] where a Small Party of Indians resided possessing some of Mr Smiths things which they restored viz—15 Beavers—2 Horses 1 Rifle & 1 Pistol—Continued our progress about 6 Miles to another Small Stream[173] also Indians residing on its banks a few Miles east of our Camp

30 Heavy Rain all Night—Sent Messengers to River *Saoustla,*[174] to have Canoes in readiness for us by the time we get there—recovered 1 Horse

31 Sultry Weather—proceeded about 8 Miles to River Saoustla—firing of Guns as a signal, agreed upon with the Indians we sent ahead, brought them to us with Canoes—and we ascended to the first Fork on the left hand and Encamped, to the Eastwards of our position Stood two Indian dwellings, sent some of our followers to endeavour and obtain information.

Novr 1 Heavy Rain all day—proceeded up the Main Channel and took our Station for the Night Opposite the first Chief Village—

2 Rainy Weather Still Continued.—the Indians of this Village restored—4 Beavers—1 Musket Barrel—1 Blanket 1 Woolen Shirt—Some Beads—3 horses—they informed us of some friendly Indians having taken forward to the *Kellymoux*[175] 12 horses and ten Beavers belonging to Mr Smith, George came to us.

3 Fine Weather—Engaged Indians to take the horses down by the woods—hardly had they Started when information was brought to us that one of the Animals was too much reduced to proceed, Sent a confidential person to see and finding the case to be correct put an end to the Animals sufferings—the Indians engaged by us went forward with two and an[other] got on the Way made up the Number to three—Encamped at our Station of the 31 Ul°—traded a few Bv^{r} Skins

4 Fine Weather—the horses arrived late in the Evening —from the Indians about here we recovered 2 Kettles—3 Saddles—

5 Cloudy Weather—Proceeded on our return, reached our first Encampment leaving the Umpqua River and Stoped for the Night—

6 Heavy Wind & Rain—reached the Spot we left our Canoes, found every thing Safe also the furs & left in charge of the Natives going forward and taken up on our return—

7 Heavy Wind and Rain—which Stoped our progress

8 The Rain Continued as yesterday and the Wind with increased violence, so that no craft could venture on the water with Safety—

NOVEM^{r} 9th Weather fine & Calm—Started by Water a few Miles Southerly—left our Crafts and proceeded by the beach about 12 Miles, to Quick Sand River,[176] found a large Party of Indians Stationed here, not usually the Case, *Nooze* an Old Acquaintance, came to us in fact he had no other alternative as his communication with his Village was cut of[f]—as we came upon them unobserved—took a position within a Short distance of the Village for the Night.—

10—Fine Weather—Our business with these Indians was over at an early hour—the following Articles were recovered—viz—3 horses—2 Mules, 7 Steel Traps, 1 Copper Covered Kettle—1 Rifle—1 Rifle Barrel—Some Beads, Books, journals & other Papers—traced back our Steps to the Entrance of the Umpqua, taking forward the horses & Mules, but had to leave them on the South Shore, the Wind and State of the Water not permitting to Swim them over the Channel safely—at Dark reach-our Camp all Safe—

11—Heavy Rain & high Wind—delayed in camp

12—In course of the last Night the Storm subsided and this Morning brought on fine Weather—a party of Men went for and brought the horses & Mules we left day before yesterday—Meanwhile Mr Smith with three Men succeeded in finding the other six horses—and at Noon moved up the River, the horses conveyed by Indians along the Margin of the River, Encamped on the Island at the Entrance of the North branch, now called *Defeat River,*[177] the Indians conveying the horses had to leave them and come to us late—

13—Fine Weather—a Party took the Woods for Game & P.[ierre] L'Etang one of the number, returned to acquaint us that they had seen two horses on the South shore of the Main River, Mr Smith with ample Means in hands, instantly went forward directed to the Spot by L Etang and found a Mule and a horse in good Condition, both brought across and had to Swim above a Mile—others were employed after the other horses and by Noon were all Collected on the East shore of the Channel, the route the Indians purpose taking by—Mr Smith proposed to accompany them conditionally that I would grant him the Assistance of two Men which I readily assented to and appointed [William] Johnston and [Amable] Quesnel to be of the Party with three Indians and himself making Six in all—

14 Fine Weather—Agreeable to the Settlement of yesterday, Mr Smith and those appointed to accompany him Started early and Shaped their Course Eastward along the Banks of the River,—having taken in a Small Supply of Venison, we proceeded up the River and occupied the Spot we did leaving the Verveau opposite to the Village, the Articles we left here on our way down found Safe and in good Order—

15 Heavy Rain, that lasted all day—Kept Camp

16 Heavy Rain Continued all day—proceeded up the River to our Camp above *Verveau,* stoped at the Village for our horse Agres[178] left en Passent [*sic*], found all in high order

17 Rain falling in torrents all day—In the afternoon La fram[boise] with Some Men went to the Village and received from Joe[179] and his Brother, 72 Large Beavers and 16 Large Land Otters, unpaid for—

18 In the Night the Rain ceased and with the rising Sun came on fine Weather.—Mr Smith & Suite joined us having been obliged to leave the horses to be taken forward by the Indians, after they have recruited which their exhausted State loudly call for—Johnston, in cutting brush wood to clear the track got a bad cut on the leg bone—Quesnel hurt in the Knee badly from the Kick of a horse.—the Indians that accompanied Mr Smith behaved much to his Satisfaction and promise fairly to render the horses up Safely and a remuneration is held out to them and Stipulated, the Articles they are to receive—Our horses left here are dispersed but four can be found in this Vicinity.—the others are about River la Biche—

19 Light Rain.—Started P. L.Etang & three Men, *En Canot* to convey the furs &c to River la Biche—and by the Close of day, we reached that place by land and had a sight of our horses Grasing in the hills

Indians informed us that our horse Guard got intimidated at flying reports circulated by Straglers, and abandoned his charge—

20 Moved to the South Bank of the River la Biche caught our horses, with the exception of two (Supposed stolen by Indians) got 2 Mules of Mr Smiths & 3 Beaver Skins —L Etang arrived and delivered his Cargo Safe—Light Rain—

21 Weather fine—rised Camp Course Easterly over the Mountain and encamped at the Forks

22 Rained all day with high Southerly Wind
Continued our progress and in good time reached our Camp, all Well—In My absence M^r Mc.Kay recovered Several Articles of Mr Smiths Property which is included in the annexed Account Showing also disbursements to Indians for Services, as in no Instance have we given property in return for any part of Mr Smiths things the Same document will point out the different Articles received and made over to the Original Owner—

23 Same Weather as yesterday—Mr Smith acquainted me with having resolved on proceeding to Fort Vancouver and to discharge such of his Men as pleased to accept their dismissal, and pay them their balance in his hands their full Amount, it seems to be his intention to give up his horses to the Company, I declined coming to terms and deferred him to a future Period to settle that and other Matters with my Senior Officer—

24 Stormy Weather not having abated, every thing is at a Stand—

25 The Stormy Weather somewhat subsided—Mr Smiths Property put across the Channel below the Forks the object is to take advantage of the Weather to proceed with all possible diligence to Fort Vancouver—La Framboise & two Men accompanies them, and it is

settled that they proceed together as a precautionary Measure—

26 Heavy Rain—People on the alert after Strayed horses—Indians seem intent on giving troube as they take horse[s] away and leave them in the field at the end of their journey—Measures must be adopted to check them in the future—

27 the stormy Weather Still prevails—and of serious inconvenience to us, as many in the Camp feel severely in consequence of it—as it renders hunting very precarious—

28 The last night remarkable for a violent Storm but as the day dawned the Weather became more Settled, and Mr Smith & La Framboise moved forward—and preparation made in the Camp to move from hence—

29 The late Rains has inundated the low Land and the different Streams overflowed their banks—

30 Rainy Weather—

DECr 1 Left Camp with 4 Men to Proceed to Fort Vancouver on the Companys business—J.[ean] B[ap]t.[iste] Perreau[l]t & L.Etand, are of the Party accompanying Me, both ailing and in need of Medical assistance about 2 P.M. we came up with Mr Smith and La Framboise on the North Bank of River La Biche—and Encamped together, Owing to the rise of the water in the Numerous Streams and the inundated State of the low land renders travelling extremely bad and very injurious to horses.—it is therefore expedient to built temporary Canoes to descend the Wullamette River to Save the furs from further injury—

2 Rain & Snow alternately during the day—proceeded forward and past Mountain La Biche at its base on the North side and put up for the Night—from Stragling Indians, the Death of *Cadonette* was intimated to us—

3 Westerly Wind & Rain, Continued our route in Company and Encamped on the Banks of Yellow River, a Branch of the Wullamette, and we fell on its Source, it is but a small Stream in Summer but now, in appearance a large River, it heads in Mountain la Biche—

4 Moved a Short distance to a More Convenient Spot on the Same Stream, to answer our Views as there is timber of Sufficient Growth for the purpose intended—Light Rain—

5 Cloudy Weather and Occasional Rain—Men employed about a Canoe—Laframboise

6.7.8.9. Visited *Charlas* and returned with two Slaves that absconded from our Camp early in October but now willing to reassume their former Situations—after they came back two More were brought in—all four deserted at the Same time, the Eight Inst—a Canoe was fit to run before the Stream and the hands were put After a Second of much larger dimensions—From Charlas Indians we received Skins in payment of a Balance due the Company, among the Number, such as belonged to Mr Smith were Selected and given up to him amounting to 23 Large Beavers and 1 large land Otter—

10 Fine Weather, with three Men & two Natives Started for the Establishment by Water—Appointed [Alexis] Obichon, horse Guard with two Native assistants at a stated Period to bring the horses to River *L'ommitomba*[180]

for the H.[onorable] H.[udson's] Bay Company
Alexr R. McLeod

Letter of Governor George Simpson to Jedediah S. Smith, Fort Vancouver, 26th December, 1828.[181]

Mr. J: S: Smith

Present

Dear Sir,

As you have had a great deal of communication with M[r] McLeod on the subject of your affairs in this quarter in the course of your late Journey to the Umpqua and as that Gentleman is now on the eve of taking his departure hence on a Voyage which may occupy him from 12 to 16 Months I consider it proper that we should come to a final understanding or Settlement on all matters relating to business while he is on the spot and in order to guard against any misapprehension that our communications thereon should be in writing instead of Verbal.—

You are aware that previous to your arrival here in the Month of Aug[t] last M[r] McLoughlin the Hon[ble] Co[y's] principal representative at this place determined on sending a party under the command of M[r] McLeod on a Trapping & Trading Expedition in a Southerly direction from hence & that the equipment of this party was nearly completed when you to our great surprise appeared at this Establish[t.]

The melancholy report you brought of the destruction of 15 Men out of your party of 19 a few days previous on your way from St Francisco to the Columbia by the Natives of the Umpqua and of the pillage of your property excited in the minds of the Gentlemen here the most lively feelings of Sympathy and comiseration and [influenced] by those feelings towards you and your unfortunate companions Dr. McLoughlin instructed M[r] McLeod to proceed with his party to the Umpqua to communicate with the Natives, to ascertain the cause of their atrocious conduct, to punish them should it have been considered expedient & found practicable and to endeavour to recover your property.

He accordingly went thither, his party consisting of 38 Servants and Indians and accompanied by you and your surviving followers.—While on the spot he learnt that the Melancholy catastrophe was occasioned by some harsh treatment on the part of your people towards the Indians who visited your Camp some of whom they said had been beaten, and one of them bound hands & feet for some very slight offence; which treatment they further said corroborated in their Minds a report that had preceded you from Indians that your party had been conducting themselves with hostility towards the different Tribes you passed in your way from the Bona Ventura (for which it appears there were some grounds) and that as a measure of Self Preservation they determined on the destruction of your party which its injudicious conduct and unguarded situation enabled those savages to accomplish with little difficulty or danger to themselves—

Mr McLeod under all circumstances found that it would be unsafe and unpolitic to take any hostile steps against the Tribe but endeavoured to recover of the property of which you had been pillaged and with some trouble and difficulty suc-
he has thus recovered consists of about 700 Beaver Skins, 39 Horses and a few other articles of little value.—

When Mr McLeod and his party took their departure Dr McLoughlin did not conceive that any inconvenience or delay would have been occasioned by their visit to the Umpqua he did not therefore intend to have made any charge against you for the Services of Mr McLeod & his party in the recovering of your property but the time occupied in visiting the different Camps on the River & Coast with that object we now find has occasioned the loss to us of the Services of this Expedition for the whole Season thereby subjecting us to an expense of exceeding £1000 independent of the loss of Profits we had reason
ceeded in getting nearly the whole of it restored—The property
to calculate from the Services of this Expedition.

Had you been in the condition of discussing terms with us, we should as a matter of course have insisted on your defraying the expences, that the recovery of your property might have occasioned to us, but you was not in that condition consequently nothing was said on the subject, and altho' we are well aware that either in Law or Equity we should be fully entitled to Salvage, we make no claim thereto, on the contrary place the property which we have recovered at your disposal without any charge or demand whatsoever

In order to suit your own convenience, you left 38 Horses at our Camp on the Umpqua which the Expedition had not the least occasion for as M^r McLeod having independent of them about 150 being more than sufficient we conceive to meet his demands; these and a few others expected to be received in order to accommodate you we are willing to take off your hands at 40/St^g p. head, which is a higher price than we ever pay for Horses and the same we charge to our Servants & Trappers but if you are not satisfied with that price, they are still quite at your disposal

In conference you have had with me both toDay and two days ago, you told me that you was desirous of taking your Furs up by Water immediately to our Establishment of Walla Walla, that there you wished us to give you Horses in exchange for those left at the Umpqua and that in the event of our complying with that wish you would leave Horses & Furs at Walla Walla while you proceeded across from thence to your Depot on Salt Lake from whence you would in the course of next Summer send for both.

In reply I now beg to state that we should consider it the height of imprudence in you to attempt going up the Columbia with only your two followers either light or with property.— We a[l]tho' perfectly acquainted with every Indian on the communication rarely venture to send a party even with Letters and with property never less than from 30 to 40 Men; such a measure on your part would therefore in our opinion be sporting with Life or courting danger to madness; which I should not consider myself justified in permitting without pointing out to yourself and followers in presence of witness's the desperate hazards you would thereby run

I should consider it equally imprudent to attempt a Journey from Walla Walla to Salt Lake on many considerations, the most prominent of which are, the great danger to be apprehended from roving War parties, your total ignorance of the Country, the difficulty you would have in finding your way across the Blue Mountains, the inexperience of your people in Snow Shoe Travelling (one of whom I believe never saw a Snow Shoe) and the danger from Starvation as it is impossible you can carry provisions such a distance and the chase in some parts of the country through which you would have to pass is at this Season even to a hunting party a very precarious means of subsistance.—In reference to your demand upon us for Horses at Walla Walla it cannot be met by any possibility as by the last advices from thence we [have] none at that Establishment and our own business in the Upper parts of the Columbia requires at least five times the number we are likely to be able to collect in the course of next Season

You are well aware that we have already experienced much inconvenience incurred many sacrifices, and exposed the Concern to heavy loss, through our anxious desire to relieve, assist and accommodate you we are willing nevertheless to do whatever else we can without subjecting ourselves to further loss or expense in order to meet your wishes, I shall now suggest what I conceive to be the safest course you can pursue and the most eligible plan you can adopt.

Your Beaver which is of very bad quality the worst indeed I ever saw, having in the first instance been very badly dressed & since then exposed to every storm of Rain that has fallen between the Month of April & the 22nd. Inst. consequently in the very worst state of Damage, I am willing to take off your hands at 3 Dollars p Skin payable by Bill at 30 d/ sight on Canada, which I conceive to be their full value at this place, and your Horses I will take at £2 St^{g} p Head payable in like manner But if these terms are not satisfactory to you the Furs may be left here until you have an opportunity of removing them & the Horses are at your disposal where you left them

In either case yourself and followers shall be made welcome to a continuance of our hospitality while you choose to remain at our Establishment—and if agreeable you shall be allowed a passage free of expense to Red River Settlement with me in the course of next Spring & Summer from whence you can proceed to St. Louis by Pra[i]rie du Chien or you may accompany our Snake Country Expedition next Autumn by which means you will in all probability have a safe escort until you fall in with your people at or in the neighborhood of Salt Lake

After you have fully considered these suggestions which are dictated by the best feelings towards you and an intense anxiety for the safety or yourself followers & property I have to request the favor of a reply thereon in Writing previous to M^{r} McLeod's departure &

I remain

Dear Sir

Yo: Mo: Ob^{t}. Ser^{t}.

Geo: Simpson.

Letter of Governor George Simpson to Jedediah S. Smith, Fort Vancouver, 29th December, 1828.[182]

Mr. J: S: Smith
Present

Fort Vancouver, Columbia River
29th. Decr 1828

Dear Sir,

In reference to your valued communication of 26th Inst., and to our subsequent conferences I beg it to be distinctly understood that we do not lay claim to, nor can we receive any remuneration for the Services we have rendered you, any indemnification for the losses we have sustained in assisting you, nor any Salvage for the property we have recovered for you, as, whatsoever we have done for you was induced by feelings of benevolence and humanity alone, to which your distressed situation after your late providential escape & the lamentable & melancholy fate of your unfortunate companions gave you every title at our hands.—And I beg to assure you that the satisfaction we derive from these good offices, will repay the Honble Hudsons Bay Comp^y amply for any loss or inconvenience in rendering them

I am exceedingly happy that you have consented to abandon the very hazardous Journey you contemplated and that you have allowed yourself to be influenced by my advice to pursue the safer yet more circuitous route by Red River, which notwithstanding the increased distance, will in point of time be the shortest, as thereby you will baring accidents be at St. Louis in the month of July next

With regard to your property, we are willing in order to relieve you from all further concern respecting it, to take it off your hands, at what we consider to be its utmost value here say Horses at 40/. each which you know to be a higher price than we ever pay for any, and Beaver at 3$ p^r Skin Land Otters at

2\$ p^{r} Skin and Sea Otters at 10\$ p^{r} Skin which from their damaged state I conceive to be their utmost value here, fully as much as they will net to us in England, and after making a fair deduction for risk and expence of transport hence to St Louis, more than they would yield you if taken to and sold in the States

But if these prices be not satisfactory to you, and that you prefer leaving your property here until a favourable opportunity should present itself for removing it, we shall with pleasure retain it for you, and deliver it when and to whom you may direct

With Esteem
I remain
Dear Sir
Yo: Mo: Ob^{t}: Ser^{t}:
Geo: Simpson

Extracts from Governor George Simpson's Report to the Governor and Committee of the Hudson's Bay Company, London, dated Fort Vancouver, 1st March, 1829.[183]

Fort Vancouver, 1st March, 1829

Hon[ble.] Sirs

..........There was an American party in the Snake Country as long ago as 1809 or 1810, who established themselves at a place called (after their Leader) Henrys Forks; but who remained only one Season, finding themselves in danger of being cut off by War parties.—Their next visit, was in 1824, when Gen[l] (a Militia Gen[l]) Ashley of St. Louis, (who notwithstanding his dignified title has had a number of ups and downs in life having been a Farmer a Shopkeeper, a Miner and latterly an Indian Trader) fitted out a large party of Trappers & Servants. Smith the conductor of one of his parties, joined our Expedition in the Autumn of 1824, and passed part of the following Winter at the Flat Head Post, taking the benefit of M[r] Ogden's protection from thence to the Snake Country where they parted, and immediately afterwards in return for our hospitality and protection, [Johnson] Gardner the leader of another of Ashleys detachments, on falling in with M[r] Ogden, laid his plans to decoy our Trappers and break up our Expedition, in which he succeeded.—

Ashley's returns that Year amounted to between 5 & 6000 Beaver, a great part of which however was taken out of what is called the "Black feet Country", about the head Waters of the Missouri.—In 1825/26 Ashleys party was made up by our Deserters, and a re-enforcement from St. Louis, to about 100 Men, who hunted in small parties all over the Snake Country, and about the Eastern skirts of the Mountains, and collected about the same quantity of Beaver; when, he retired from the business with a fortune, which in Dollars sounded large in the

United States, and resumed his Shopkeeping concerns in St. Louis: but the fortune in question, was entirely nominal as the profits arising from the two prosperous Years on the West side barely covered the losses sustained during the two preceding Years on the East side the Mountains; the fact therefore is, that Ashley gained merely a little eclat by his trapping speculations, notwithstanding all the bombast that appeared in the American NewsPapers of 1824, 1825 & 1826 in regard to their "enterprizing Countryman".—The Trapping business was then taken up, by three of "the Gen^ls" late conductors; men who had formerly been practical trappers, but who all at once promoted themselves to the Travelling title of Captains, while their Mercantile operations were conducted under the firm of Smith Jackson & Siblit. Their first year was prosperous having collected from 5 to 6000 Beaver; but since then, they have been very unfortunate.—With regard to Jackson & Siblit, we learn that they had several parties scattered about the Snake Country some of whom Mr Ogden saw, but they complained of the poverty of the Country, had lost the greater part of their Horses without which they could do little good, and one of their parties consisting of [Samuel] Tulloch and Eleven Men we last Autumn understood had been cut off by the Blackfeet.—Jackson, accompanied by a Clerk Fitzpatrick, and a Major Pilcher with a Clerk Gardner & 40 Trappers, was the band alluded to as having visited the Flat head Post last Winter; they had very few Skins, and of those few, about half fell into our hands in exchange for some necesary supplies.—Pilcher, who made his first appearance on this side of the Mountains last Summer, is the head of a Trading Association called the "Missouri Fur Co^y" of St. Louis, which failed in the year 1825 and 'tis probable the same fate awaits his present concern as it must have been in a desperate state indeed, when the head thereof, could not find better employment for himself and followers than watching the Flat Head Camp.—"The Major" and Smith Jackson

& Siblit, are in hot opposition to each other, and both court our protection and countenance, while we contrive to profit by their strife. Pilcher, has made a formal tender of his Services to the Hon[ble] Co[y] by Letter addressed to me, but I have rejected his strange proposition, in terms which will shew the American Gov[t] if necessary, that we pay due respect to their rights of Territory & Trade, see correspondence.—Smith, the head of the Firm of Smith Jackson & Siblit who now enjoys our hospitality and protection, (and whom I have already noticed as having been with M[r] Ogden at the Flat head Post) has been truly unfortunate, and as the circumstances which placed him here, may become a subject of future misrepresentation and enquiry, I shall now detail them, principally from his own report, for Your Honors information.—

In the Summer of 1826 he started with a party from Salt Lake, for the purpose of Trapping the Rio Colorado, where he got some Beaver; but falling short of Ammunition and other supplies, he proceeded down that Stream until he found the Macabas Tribe, who received him kindly & conducted him to St. Gabriel, where he and his party were treated as Prisoners;[184] but liberated under a promise that he would quit the Spanish territory; from thence, he proceeded along the Coast to St. Francisco, where he received a few supplies in exchange for Beaver.—The Spaniards throughout, looked upon him with much suspicion, he underwent a thousand cross-examinations but they never could believe that his sole object was to hunt Beaver, an Animal they scarcely knew by name, altho' some of the Rivers within a few Miles of their Settlements, abounded therewith: they however permitted him to depart, and as he required additional Strength and supplies, to enable him to hunt this Country, he left Eleven Men to trap in the Waters of the "Buona Ventura," and in Spring 1827, he with a couple of Men, crossed a Sandy desert of considerable extent (in which he suffered greatly from a scarcity of Water) to their Depot at

Salt Lake, where he arrived in the Month of July.—After having rested a few Days there, he started again with about 20 Men, following his former route by the Rio Colorado, (as he would not attempt to recross the Sandy desert) and as formerly fell in with the Macabas Indians, some of whom it appears had been severely punished by the Spaniards for conducting him on the former occasion to St. Gabriel, and who had instructions to allow no Strangers to pass by that route again.—

These Indians at first received him kindly as before, but soon took an opportunity when off their guard, and while Swimming across the River, to attack the party, Ten of whom they succeeded in destroying.—Smith and his surviving followers, however, got down to St. Gabriel a second time,[185] where they were again made Prisoner and detained for several Weeks, but at length permitted to go to St. Francisco where their second appearance excited more astonishment and alarm if possible than before. Here, they were again confined, and examined in all manner of ways, and it was here that Lieut. Simpson saw him in December 1827: he was however permitted again to depart early in January 1828, after having purchased about 300 Horses with the proceeds of his Beaver.—These Horses cost about 10$[186] each, and had they reached the Depot would have met a ready Sale to his Free Trappers at 50$ each.—

Smith's party now united with those left hunting when he crossed the Sandy desert in Spring 1827, amounted to 19 in all.—From St. Francisco they took a Northerly course along the North branch of the Buona Ventura, found the River well stocked with Beaver, but only hunted while it was necessary to rest their Horses; his object being to push on to the Depot with the Horses, and conduct a large body of Trappers back in order to Scour the country, they however caught about a Thousand Beaver.—

Their object in taking this Northerly course, was to fall

upon the Wilhamot [Willamette River], and proceed either by the Columbia or across country from thence to Salt Lake, being desirous of avoiding the circuitous route by the Rio Colorado, and unwilling to attempt cutting across the Sandy desert: but they found the country much more rugged & Mountainous than they expected, and were obliged to pass round by the Coast. In the course of this Journey, they repeatedly fell in with Indians whom (we learn from other Tribes) they regarded as Enemies.—At length they reached the Umpqua River in July last, distant from hence about 150 Miles, where they encamped to recruit their Horses.—While at this encamp[t], Smith with a couple of Men and an Indian, went in search of a favorable route for their Horses, leaving 16 Men in the Camp which was surrounded by a large body of Indians, who appeared to be on a friendly visit to them, but who at a given Signal attacked the camp, and destroyed the whole party, except one Man who saved himself by darting into the Woods.—Smith, on his return in a Small canoe fortunately discovered before landing, that the Massacre had taken place, otherwise he would have shared the fate of his comrades; but with his two Men paddled to the opposite of the River, and saved themselves by flight into the Woods.—The Man who escaped the dreadful camp scene, fell into the hands of Friendly Indians, by whom he was conducted to this Establishment, and Smith and his two followers made their appearance here a few Days afterwards, on the 10th of August, where they were received with every kindness and hospitality.—

At the time of Smith's arrival, Chief Factor M[c] Loughlin was fitting out a trapping party to hunt under the direction of Chief Trader M[c]Leod in a Southerly direction from hence along the Coast.—This party was to start a few Days afterwards, and with the double object of recovering Smiths property all of which fell into the hands of the Indians, and of enquiring into the cause of, and punishing those who were concerned in the horrible outrage if found practicable and considered expedient, M^r^. McLeod was directed to conduct his Expedition by the Umpqua, which he did accompanied by Smith.—

On arrival there M^r^ McLeod Summoned the principal Chief and his followers to the Camp, which they obeyed, and in answer to the queries put to them as to the cause of the Massacre, they said, that previous to Smith's arrival they had notice of the approach of his party, from some of the Tribes he had passed, with intimation that they were Enemies destroying all the Natives that came within their reach. That this information was in some degree confirmed by their severely beating and binding the hands and feet of one of their own Tribe who had pilfered an axe, (a very slight offence in their estimation). That they declared themselves to be people of a different Nation from us, and our Enemies, and therefore intended to drive us from the Columbia where we were intruders on their Territory.—

These circumstances, they said, induced them to look on the party with suspicion, but they had not formed any plan of destruction, until one of them "Rogers" a Clerk, in Smith's absence, attempted to force a Woman into his Tent, whose Brother was knocked down by Rogers while endeavouring to protect her; upon which, seeing the opportunity favorable, as some of the people were asleep, others Eating and none on their guard, they rose in a body and dispatched the whole party except the man who fled.[187]

Some parts of this Statement, Smith denies; but the whole story is well told, and carries the probability of truth along with it.—M^r^ M[c]Leod might have taken the lives of several of the Murderers; but had he done so, it would have involved us in eternal Warfare with a very Numerous and powerful Nation, with whom we have been on Friendly terms for several Years, whose trade is of importance to us and in whose power our Trapping and Trading parties would frequently be; he therefore considered it prudent as regarded our own safety, and politic as regarded the interests of the Service, to abstain from violence; but took much trouble in recovering such part of the property as was within a convenient reach.—The property thus recovered through M[r] M[c]Leods exertion and influence, was from 7 to 800 Beaver & Otter Skins in a very damaged state, 40 Horses, and a few other articles of little value.—This business occupied much time, from the latter end of August until the Month of December, and I am concerned to say has occasioned the loss of the Services of this Expedition for the Season.—M[r] M[c]Leod came back to this Establish[t], for further instructions, accompanied by Smith, leaving his Expedition at the Umpqua, and after remaining here a few Days started afresh with directions to proceed to the Southward, as p copy of my Letter of instructions of 29th December herewith transmitted.—

Smith soon afterwards intimated his intention of proceeding to his Depot at Salt Lake, but the undertaking appeared to me so hazardous, that I remonstrated against it, and he fell in with a proposition made by me, to take his Furs and Horses at a given price as p account, and to give him a passage to Red River in Spring, from whence he can with little risk push his way to St. Louis.—All my business communications with this person, have been by Letter, under Dates 26th & 29th Decem[r], copies of which with his answers are herewith transmitted; and altho' we have sustained considerable loss, by our endeavours to be of Service to him, we have no doubt that your Honors will approve the feeling by which we were actuated and the course we pursued in reference to this melancholy affair.

We learn from our American visitant Smith, that the flattering reports which reached St. Louis of the Wilhamot Country, as a field for Agricultural speculation, had induced many people in the States to direct their attention to that quarter; but he has on his present journey, discovered difficulties which never occured to their Minds, and which are likely to deter his countrymen from attempting that enterprize.—In the American Charts this River, (the Wilhamot or Moltnomah) is laid down, as taking its rise in the Rocky Mountains, (indeed Mr [Richard] Rush in his official correspondence with President Adams on the subject of a boundary line distinctly says so) and the opinion was, that it would merely be necessary for Settlers with their Horses, Cattle, Agricultural implements &c. &c to get (by the Main communication from St. Louis to Sta. Fee) to the height of Land in about Lot 38, there to embark on large Rafts & Batteaux and glide down current about 800 or 1000 Miles at their ease to this "Land of Promise".—But it now turns out, that the Sources of the Wilhamot are not 150 Miles distant from Fort Vancouver, in Mountains which even Hunters cannot attempt to pass, beyond which, is a Sandy desert of about 200 Miles, likewise impassable, and from thence a rugged barren country of great extent, without Animals, where Smith and his party were nearly starved to Death. And the other route by Louis's River, Settlers could never think of attempting.—So that I am of opinion, we have little to apprehend from Settlers in this quarter, and from Indian Traders nothing; as none, except large capitalists could attempt it, and this attempt would cost a heavy Sum of Money, of which they could never recover much.—This they are well aware of, therefore as regards formidable opposition, I feel perfectly at ease unless the all grasping policy of the American Government, should induce it, to embark some of its National Wealth, in furtherance of the object.—

I have the honor to be
Honble Sirs
Your Mo: Obedt. Humb. St.
Geo. Simpson

To the Govr. Depty Govr & Committee
of the Honble Hudsons Bay Coy.
London.

Betsey Smith, sister of Jedediah

Sally, eldest sister of Jedediah Smith

Eunice Smith, called the Great White Medicine Woman by the Indians. Sister of Jedediah.

Benjamin G. Paddock Smith, who preserved his brother's transcript **Journal.**

Letter of Governor and Committee of Company Directors to Chief Factor John McLoughlin, dated London, 28th October, 1829.[188]

We are much gratified to hear that every hospitable attention and assistance were offered to Mr. Smith the American and his Companions in distress after the horrible massacre of his party by the natives of the Umpqua, and from the humane feeling you have already manifested it is scarcely necessary to desire, that you will on all occasions render any protection in your power to Americans, Russians, or any other strangers who may be in the Country against the treachery or violence of the natives whatever may be the objects of the visits of such strangers, be they competitors in trade or otherwise, as all feeling of self interest must be laid aside when we can relieve or assist our fellow creatures

As the foregoing letters reveal, when Jedediah Smith returned from the McLeod expedition he found George Simpson,[189] Governor-in-Chief of the Hudson's Bay Company territories, visiting Fort Vancouver. The Governor had arrived there on October 25, in the course of a tour of the Company posts. He brought Jedediah's property, giving him a draft on Canada for approximately $2,600.

On March 12, 1829, Smith and Arthur Black traveled up the Columbia, bound circuitously for Jackson's Lake, where, nearly two years before, he had agreed to meet William Sublette and the main body of trappers.

He traveled far up the Columbia to Fort Colville and Kettle Falls in the northeastern corner of what is now the State of Washington, turned to the southeast along Clark's Fork, and in the Kootenais country came upon the company of David E. Jackson, who had been looking for him in the northwestern part of what is now Montana.

With Jackson and his men, Smith went down to Henry's Fork of the Snake River, where they joined Sublette. The partners, reunited after two years in which Smith was adventuring, all proceeded to Pierre's Hole, a celebrated haven nearby in the wilderness.

Meanwhile, as a result of the California expedition, rumors were flying in Mexico. It was said that the United States intended trying to seize the port of San Francisco.

On Christmas Day, in the year 1830, Sublette set out for St. Louis with a revolutionary plan in his head. He was going to fetch wagons over the prairie into the mountain country, traversing ground that had never known the touch of a wheel save those of the cannon which Smith found at Bear Lake when he returned from his first journey to California.

The wagon train arrived at the Wind River rendezvous of 1830 in July of that year. Smith, Jackson & Sublette sold out to Thomas Fitzpatrick, Milton G. Sublette, Henry Fraeb, Jean

Nelson J. Smith, brother of Jedediah

Ira G. Smith, brother of Jedediah

Peter Smith, brother of Jedediah, as he looked on the trail

Peter Smith in civil life

One of Jedediah Smith's pistols, recovered from the Comanche Indians by comancheros (Mexican traders with the Indians) and by them taken to Santa Fé, where it was surrendered to Peter and Austin Smith. **Historical Collection, Security-First National Bank of Los Angeles.**

Baptiste Gervais and James Bridger, who organized as the Rocky Mountain Fur Company. The retiring mountaineers arrived with their wagons at St. Louis early in October, bringing in "a large quantity of furs . . . richly rewarded for their perils and enterprise."[190]

Smith chose a farm in Ohio; made gifts to relatives and friends; engaged Samuel Parkman to copy his notes for publication; drew, with the collaboration of Jackson and Sublette "a new, large and beautiful map, in which are embodied all that is correct of preceding maps, the known tracks of former travelers, his own extensive travels, the situation and numbers of various Indian tribes, and much other valuable information";[191] and embarked two of his brothers, Peter and Austin, in the Santa Fé trade, with his former partners, Jackson and Sublette.

At the last moment Jedediah decided to accompany the caravan.

Between the Arkansas and the Cimarron Rivers, in the southwestern part of what is now Kansas, Smith, alone, was scouting for water, which had been lacking three days. He found it at length, but he was ambushed by a band of Comanche Indians, who, disregarding his invitation to go to the wagons and trade, frightened his horse by waving blankets and flashing the light from mirrors into its eyes.

The horse turned and Smith was exposed to a Comanche lance. Wounded, he fired and killed two of his attackers. The rest closed in upon him then, and, on the 27th of May in the year 1831, Jedediah Smith came to the end of his last trail.

Letter of Austin Smith.[192]

Walnut Creek on the Arkansas, 300 Mile from Settlements.

Septbr. 24th, 1831

Dear Brother—

An opportunity offers itself of writing to you from this point by some gentlemen who are anxious to reach their families.

It is my painful duty to communicate to you the death of our lamented brother, Jedediah. He was killed by the Comanche Indians on the 27th May on the Simarone River between Arkansas River and Santa Fé.

His company and Soublett's, consisting of 74 men, and animals for 22 wagons, was on the point of starving for the want of water (near four days without any). He took a due south course from the one we were travelling, which was S.W., and struck the Simarone. The Spanish traders who trade with those Indians informed me that he saw the Indians before they attacked him, but supposed there could be no possible chance of an escape. He therefore went boldly up with the hope of making peace with them, but found that [his] only chance was defense. He killed the head chief. I do suppose that then they rushed upon him like so many bloodhounds. The Spaniards say the Indians numbered from fifteen to twenty. I have his gun and pistols, got from the Indians by the traders.

Such, my dear brother, is the fate of our guardian and protector on this route; him who had gone through so many dangers, so many privations; and almost at the time when he had reached the goal of his enterprise, to be thus torn from us is lamentable indeed. But let us not grieve too much, for he confided in a wise and in a powerful Being.

Peter left Santa Fé on the 29th of August in good health for California with a party to purchase mules. They have taken part of our merchandise.

I will you would meet me at St. Louis about the 25th of

October, as I shall be there about that time. I shall make calculations to stay in St. Louis this winter, when we can have an interchange of sentiments. I will give you further particulars of what we can do with Peter. Meanwhile, I am, dear Brother, yours forever.

N.B. The Spanish traders say that the Indians succeeded in alarming the horse he was riding so as to get his back to them, which, when effected, they then forced on him and wounded him in the shoulder. He then faced them and killed the Chief.

Letter of Solomon A. Simons, Ashtabula, Ohio, to Ralph Smith, Mohican, Wayne County, Ohio.[193]

Dear Brother,

Having received a letter from Austin Smith I think it my duty to write to you [and] let you know some of the contents of sd. letter; it is some grievous. Austin states that J.S. Smith was killed by the Indians the 27th day of May, 1831, three hundred miles from the settlements and that he went to the river to get water, was attacked by fifteen or twenty Indians. They did not fire at him so long as they were face to face, but they scart his horse and then wounded him in the shoulder, rushed upon him to despatch him. This is all that he wrote save that he and Peter should be at St. Louis the first of December and then he would write the particulars concerning the death of Brother Jedediah S. Smith.

I carried the letter to Father Smith last Thursday. He seemed to be more resigned than I expected. We are all in good health excepting my side is not well, but I can work some.

I have written to Austin to know whether he wants any assistance to settle the estate of J.S. Smith. Now I wish you to write to me whether you shall go to St. Louis this fall. If you think that you shall if you will write to me the time that you will start I will go with you. If you think it best I will come to your House the time you say if you want my company. If you do not go I wish you to write me on the receipt of this, to satisfy father and the rest of us concerning the estate of Jedediah's property; whether he made his will or not. Probably you know more than any of us about it; therefore father wishes to know. So I must end by subscribing myself affectionately, this 23rd of Oct., 1831.

Last Will and Testament of Jedediah Smith

HE Last Will and testament of Jedediah S. Smith made in the County of Lafayette and State of Missouri this Thirty first day of April in the year of our Lord Eighteen hundred and Thirty one, Witnesseth as follows:

1st It is my will and desire that my accounts be settled and that my debts be paid

2nd It is my express will and desire that my honored Father Jedediah Smith receive from my property annually during his life the sum of two Hundred Dollars.

3d After the arrangement of my affairs, it is my will and desire that all my property real and personal saving the Le[g]acy to my father be divided in equal parts to my Brothers and Sisters whose names are as follows[:] Ralph Smith[194] Austin Smith[195] Peter Smith[196] Ira G Smith[197] Benjn G P Smith[198] Nelson J Smith[199] Sally Jones[200] Betsey Davis[201] Eunice Simonds.[202]

4th In the division of my property, it is my wish that the ballance of my account against my brother Ralph Smith be considered as an advance made to him and of course deducted from his share.[203]

Lastly I do hereby constitute and appoint my particular and confidential friend William H. Ashley Executor on this my last will and testament.

In testimony whereof I have hereunto set my hand and Seal the day and year as above written.

JEDEDIAH S. SMITH

The above mentioned Jedediah S. Smith on the day and year aforesaid and in our presence signed the above instrument of writing and we the subscribers in the presence of the said Jedediah S. Smith and of each other have signed our names at Lafayette County in the State of Missouri the day and year as above written

SAMUEL PARKMAN
JONATHAN T. WARNER

Notes

Notes

1. Evidence in Smith's own words that he could not have been a guide on the Santa Fé trail in 1822, as affirmed by a grandnephew, Ezra Delos Smith, in *Kans. Hist. Collections,* 1911-1912, xii, 256. E. D. Smith's inquiries resulted in the finding, by the Missouri Historical Society, of the diary of Harrison Rogers, Jedediah's clerk in 1826-1828, and of some of the Mss. in the Kansas Historical Society collection; but I think his account of journeys of Jedediah Smith prior to 1823 a confused recollection of family tales heard by E. D. Smith in his childhood. The *Journal* reveals that Jedediah Smith was unfamiliar with the trans-Missouri buffalo and with organized trapping prior to 1822. Nevertheless, E. D. Smith has never received the credit due him for successful research on his distinguished kinsman.

2. William H. Ashley was in turn captain, colonel and general of the Missouri militia; later lieutenant-governor of the State. On February 13, 1822, he advertised in the *St. Louis Republican* for one hundred enterprising young men to ascend the Missouri River to its source, there to be employed one, two or three years." He engaged in the fur trade as a partner of Major Andrew Henry. Though he had some reverses, in 1826 he retired from the Rocky Mountains with a small fortune, which was increased later by supplying trade goods to Smith, Jackson & Sublette, his successors. At the death of Jedediah Smith, Ashley became administrator of estate, but Ira G. Smith, brother of Jedediah, supplanted him after a court action.

3. Andrew Henry joined the Missouri Fur Company in 1809 and was an experienced mountain man in 1822 when he became General Asyley's partner. He built a post on a tributary of the Snake River, and another at the junction of the Yellowstone and the Missouri. After two years as partner of Ashley he withdrew from the mountains.

4. Daniel S. D. Moore was deputy clerk of the Circuit Court at St. Louis in 1821.

5. On the inside of the front cover of the *Journal* appears a note: "General Ashley, I have left the commencement of your business to fill in as you are so much better informed of the circumstances. If it is your choice you can do it as if it had been done by me, which from our conversation I suppose you would prefer. The corresponding space I would suggest as being the most appropriate for giving the origin of your business in my words; the chain of narrative would be unbroken." The space allotted to the general remained blank.

6. Sniabar Creek.

7. Fort Atkinson, located where now stands the town of Fort Calhoun, about 25 miles above the modern Council Bluffs. In 1822 Lieutenant-

Colonel Henry Leavenworth, Sixth Infantry, was in command, except during a few weeks in September and October, when Colonel Henry Atkinson was in charge, and in November, when Major Alexander Cummings commanded.

8. Variant spelling of Omahas and Poncas.

9. Aricaras, Arickaras, etc., commonly called Rees.

9a. Words crossed out in manuscript. Similar deletions will be noted. All other material in brackets is supplied by the editor.

9b. Deleted.

9c. Deleted.

9d. Deleted.

9e. Deleted.

9f. Deleted.

10. Henry's Fort was at the mouth of the Yellowstone River.

10a. Deleted.

11. Names of this party do not appear in the transcript *Journal*. Mr. Chapman is probably the A. Chapman who with E. More was killed by the Aricaras on the Platte River, in 1824, as noted by Smith, Jackson and Sublette in a chart containing the names of "persons killed belonging to the parties of William H. Ashley and Smith, Jackson & Sublette," the year of death, by whom killed, etc. This chart is now among the Kansas Historical Society *Mss.*, Topeka.

11a. Deleted.

11.b. Deleted.

12. Here the first part of the *Journal* ends.

13. James Clyman, was born in Virginia in 1792. In 1823, after some experience in surveying, he entered the service of Ashley. In the Rocky Mountains he served as a leader of trapping parties, but he did not appear prominently in Western literature until 1928, the year Charles L. Camp edited for the California Historical Society *James Clyman, American Frontiersman, 1792-1881,* (San Francisco); including Clyman's record of his first experience in the mountains, obtained from a notebook in the Draper *Collections* of the Wisconsin Historical Society. A set of Clyman's notebooks was preserved by a grandson, Wilber Laman Tallman, Napa, California, and by him deposited in the Huntington Library, San Marino, California. Clyman, Wisconsin, is said to have been named for this pioneer. He is vaguely recalled by reminiscents as one Clymer or Clement. Late in life he settled at Napa, and there in 1881 he died.

14. Thomas Fitzpatrick was born in County Cavan, Ireland. In the mountain country he was known variously as Broken Hand, Withered Hand and Bad Hand, because of an accident which maimed him. In later years he was a valuable Indian agent. His letters reveal exceptional ability. His biography recently was presented by LeRoy R. Hafen and

W. J. Ghent in *Broken Hand, the Life Story of Thomas Fitzpatrick, Chief of the Mountain Men,* Denver, 1929.

15. William Sublette was born in 1799 of Kentucky stock. Ancestors on both sides were celebrated in frontier history and he himself was one of several brothers prominent in the fur trade. In 1826 he became a partner of Jedediah Smith and David E. Jackson in the firm of Smith, Jackson & Sublette. In 1832 he entered partnership with Robert Campbell, also a close friend of Smith. Sublette died in 1845, while on his way to Washington, D. C.

16. Correspondence dealing with the Leavenworth expedition is available in the *South Dakota Historical Collection,* Pierre, 1902, vii.

17. Clyman, *op. cit.,* 25.

18. John S. Robb, writing as "Solitaire" in the *St. Louis Weekly Reveille,* March 1, 1847, said Major Fitzpatrick discovered the South Pass in 1824, Smith having been left behind because of the mauling by the grizzly bear. Nevertheless a first-hand witness, Clyman, makes it clear that Smith soon was back on his horse and that he led the party through the mountains. Étienne Provot was reputed to have crossed the South Pass in the fall of 1823, but no proof is available, and critics cite evidence in disproof.

19. A letter written by Ramsay Crooks to the *Detroit Free Press,* in June, 1856, is considered evidence that a group of the founders of Astoria, Oregon, crossed the South Pass from west to east. Some commentators argue that they actually crossed some distance from the gap now known as the South Pass.

20. Ross took the Americans to be "spies rather than trappers". *Journal of the Snake River Expedition, 1824,* Ore. Hist. Soc. *Quarterly,* xiv, 385.

21. Ashley in letter to General Atkinson, December 1, 1825. Missouri Hist. Soc. *Mss.*

22. Bonner, T. D., in *The Life and Adventures of James P. Beckwourth, Mountaineer, Scout and Pioneer, and Chief of the Crow Nation of Indians,* New York, 1856. Beckwourth, or Beckwith, as he was earlier known, was an employé of Ashley.

23. Little is known of David E. Jackson, save that he was in the mountains continuously from 1823 until he, Smith and Sublette sold out and returned to St. Louis in 1830; that he embarked on the Santa Fé trail with his partners in 1831, and, after Smith's death, made his way to California. He was among the striking figures of the trap trail and had he left any writings would probably be as well known to the public as James Bridger and other scouts. Jackson Lake in Teton County, Wyoming, was named for him .

24. Jedediah Smith wrote a brief report of his first southwest ex-

pedition in a letter to General William Clark, Superintendent of Indian Affairs, dated July 17, 1827, at Little Lake of Bear River. This manuscript is in the *Letter Book* of the Superintendent of Indian Affairs, 1830-1832, in the Kans. Hist. Soc. collection, Topeka. A copy is in the files of the Bureau of Indian Affairs, Washington, D. C., and it is reprinted in Dale, Harrison C., *The Ashley-Smith Explorations and the Discovery of a Central Route to California, 1822-1829,* Cleveland, 1918, 186-193; in Hist. Soc. of So. Cal, *Annual Publications,* 1912-1913, ix, 200-203; Merriam, C. Hart, *"Earliest Crossing of the Deserts of Utah and Nevada to Southern California: Route of Jedediah S. Smith in 1826,"* Cal. Hist. Soc. *Quarterly,* January, 1924, 228 *et seq.;* and elsewhere. The *Journal* record of the second trip to California by Smith recalls events of the first.

25. Harrison Rogers, Louis Pombert, John Reubascan, John Hanna, Martin McCoy, Abraham LaPlant, Peter Ranne, Emanuel Lazarus, John Gaither, Daniel Ferguson, James Read, John Wilson, Arthur Black, Silas Gobel and Robert Evans. Two others, Manuel Eustavan and Neppasang started with the company, but they disappeared from the record.

26. Lost, Adams and Inconstant Rivers, as named by Smith, appear on the map in *Synopsis of the Indian Tribes Within the United States East of the Rocky Mountains and in the British and Russian Possessions of North America,* American Antiquarian Society *Transactions and Collections,* Cambridge, 1836, by Albert Gallatin. Gallatin received from General Ashley information based on the notes of Jedediah Smith, but he misidentified the Smith route to the Virgin with the Ashley route down the Green. Captain Benjamin L. E. Bonneville's map, (1837), shows the Smith influence. The map of the United States, (1839), by David H. Burr, geographer to the House of Representatives, is based on superior information. Gallatin, Bonneville and Burr all chart Lost River as flowing westward into a small lake; doubtless Smith learned from Indians that the stream was "lost" in a desert basin. To the superficial observer, the body of water indicated on these maps would appear to be Little Salt Lake, but Dr. A. M. Woodbury of the University of Utah, vice-president of the Mount Zion Mountaineers, H. L. Reid, ranger-naturalist at Zion National Park, and William W. Seegmiller, a stockman thoroughly familiar with the topography of Southern Utah, are united in giving me their opinion that Lost River can be no other than the Beaver. Smith gave the name Inconstant to the Mojave River because, presumably, then as now it appeared and disappeared erratically in its course.

27. Smith to Clark; see note 24.

28. The Rogers diary, reprinted in Dale's *Ashley-Smith Explorations,* give details of the party's stay at Mission San Gabriel.

29. Smith in 1826 struck two important trails: the Utah route of Fathers Sylvestre Vélez de Escalante and Francisco Atanasio Dominguez,

who in 1775 began a 2,000-mile journey through New Mexico, Arizona and Utah, accompanied by Don Piedra de Miera, a map-maker and astronomical observer, two other dons, and half-breeds and Indians; and that of Father Francisco Hermenegildo Tomás Garcés, who in 1776 crossed the Mojave Desert from the Colorado River, alone save for Indian guides. Going to the San Joaquin Valley in 1827 Smith struck the route of the soldier, Pedro Fages, taken in 1772. One of the reasons why the Spanish felt security from invasion of the desert from the east was this: in 1810 Mojave Indians stole twenty-five horses, but all the animals except one died of thirst in crossing the desert; the last died the day after arrival at the Mojave villages.

30. Rogers *Ms.*, Miss. Hist. Soc., St. Louis. Father José Bernardo Sánchez came to Mexico from Robledillo, Spain, in 1803. He was at California's Mission San Diego from 1804 to 1820. After spending a year at Mission Purísima he went to San Gabriel, where he remained until death in 1833.

31. José María de Echeandia was Governor and *comandante general* of California from November, 1825, to the end of January, 1831. He had been a lieutenant-colonel of Mexican engineers. He returned to Mexico in 1833. Echeandia was a zealous Republican, and somewhat at odds with the Spanish friars.

32. On Dec. 30, 1826, Governor Echeandia wrote to Mexico that an American or Englishman had arrived at San Gabriel Mission in command of fourteen of his countrymen, guided from the *ranchería* of the gentile Amajavas [Mojaves] by two neophytes of that mission, who had fallen away from the Christian faith; that said commander had forty beaver skins and many traps; that he turned over without question the rifles with which his men were armed and presented the enclosed letter; that by its orders he appeared to attest, accompanied by one of his trappers; that he has shown five passports for fifty-seven persons and voluntarily presented a diary of events and an itinerary, the original of which was attached; that said individual, after several questions asked him, has not shown bad faith and is judged guilty only because he entered foreign dominions where his pass did not grant him entry, but since the territory through which he crossed is desert country, his search for game may have prevented his thinking of it [!], and for that reason he [Echeandia] does not believe his entry to be malicious; that said person has his store in Salt Lake; that stating he wished permission to return there, it was not granted, awaiting orders [from Mexico City] regarding the affair. *Department State Papers,* xix, 37-38. Digest copy, Bancroft Library, Berkeley, California.

33. On Oct. 8, 1827, "T. Galbraith" (Isaac Galbraith, member of Smith's second California expedition), asking permission to stay in California or to rejoin his leader, deposed "that Smith is not, as is believed, a

military captain, but leader of a company of beaver hunters; that he has a license to collect amphibian hides. . . . . ". *D.S.P.*, ii, 39-40, Banc. Lib.

34. William H. Cunningham arrived in California from Massachusetts, in 1826. He traded along the California coast and at one time tried to found an establishment on Catalina Island, but was ordered off by the Mexican authorities. Cunningham and five other officers of foreign ships at San Diego aided Smith's release by signing a document testifying that Smith's story was correct; that he had been obliged to enter California to get supplies. Printed in Cronise, T. F., *Natural Wealth of California,* San Francisco, 1868, 43.

35. " . . . in the *ranchería* of the gentile Muquelemes they were surrounded for battle, but the Americans forming into square bodies killed five gentiles; these [Indians] seeing they were not our soldiers, became appeased . . .". *Comandante* Ignacio Martinez to *Comandante General* Echeandia, May 21, 1827, *Archivo del Arzobispado de San Francisco,* part 1, 33: A few months before Smith's arrival the Cosemenes killed more than a score of San José neophytes, and repulsed a punitive expedition sent against them.

36. Smith wrote to Father Narcisco Durán, May 19, 1827, explaining the cause of delay. Letter in Cronise, *Natural Wealth of California;* Dale, *Ashley-Smith Explorations;* Hist. Soc. of So. Cal. *Ann. Pubs.,* 1896, iii, and elsewhere. Why Smith called the Sierra Nevada Mt. St. Joseph is not clear. It may have been suggested by the proximity of Mission San José; but if Smith came away from San Gabriel with the same exalting opinion of the character of "Father Joseph" as that expressed by Harrison Rogers, he may have named the mountain range in honor of the missionary.

37. Thomas Creek, running northeast from Deep Creek Mountains. The diary resumes with Smith near what is now Gandy, Millard County, Utah, almost on the present state line.

38. Fish Springs and vicinity.

39. The Salt Desert.

40. The Stansbury Range. Smith habitually refers to a range as "the mountain"; thus he calls the entire Sierra Nevada Mt. St. Joseph, or Mt. Joseph. Scholars have been deceived by his reference to Mt. Joseph in the letter to General Clark (see note 24), and have identified it as Lassen Peak, Mt. Shasta, etc.

40a. Deleted.

40b. Deleted.

40c. Deleted.

40d. Deleted.

40e. Deleted.

40f. Deleted.

40g. Deleted.

40h. Deleted.

41. Bannocks (Pun-naks, Ponaks, Bonaks, etc.) from the Snake River. Pahnakkee is a corruption of Panaiti, the name by which these Indians called themselves.

42. Skull Valley. Charles Kelly, of Salt Lake City, author of *Salt Desert Trails,* tells me that Moodywoc, an ancient Goshute Indian, recalls hearing his grandmother say that in her girlhood three starving men, the first whites she had ever seen, emerged from the Salt Desert.

42a. Deleted.

42b. Deleted.

42c. Deleted.

43. The shores of the lake which Smith, in 1827, called his "home in the wilderness" were "first explored and described in 1843 by Col. Frémont", according to the *International Encyclopedia,* New York, 1895, vii, 52. Smith was familiar with every part of Salt Lake when Frémont was a child. It had been circumnavigated by the Ashley men.

44. Jordan River.

41a. Deleted.

44b. Deleted.

45. Bear Lake. The *Journal* transcript leaves parts of two pages blank for notes from July 3 to July 13; these were not written.

45a. Deleted.

46. Thomas Virgin, Isaac Galbraith, John Turner, Joseph Lapoint, Toussaint Maréchal, Thomas Daw, Joseph Palmer, Charles Swift, David Cunningham, Francis Deramme, Boatswain Brown, Gregory Ortega, William Campbell, John B. Ratelle, Pale, Polite, Robiseau and Silas Gobel. Gobel was one of the two men who crossed the desert with Jedediah, from California to Salt Lake. The other, Robert Evans, did not venture on the second expedition.

46a. Deleted.

47. Sentence unfinished.

48. Sevier River. Ashley Creek on the Burr map.

49. The Santa Clara River.

50. Beaver Dam Wash.

50a. Deleted.

51. Three miles south of St. Thomas, Nevada. Cave mentioned by Smith in letter to General Clark. The cave found on the second trip is probably one of the so-called calico salt mines, worked by the Indians with crude stone hammers and picks.

51a. Deleted.

52. Savants have been puzzled by Smith's statement in his letter to General Clark that the river he called Adams turned to the southeast. Dr. M. R. Harrington, curator at the Southwest Museum, Los Angeles, has done archeological research work on the Virgin River; he believes the

explorer did not take into consideration a compass deviation of about seventeen degrees. The *Journal* details will enlighten commentators misled by the Clark letter.

53. Variant of Mojave, (Mohave, Jamajabs, Amajabas, etc.); the most numerous tribe of the Yuman family, and at one time considered the most warlike.

54. Sonora included much of what is now Arizona. Although Jedediah Smith was the first to lead a trapping expedition as far as California, by the end of 1826 virtually all the streams in northern Mexico east of the Colorado had been visited by beaver hunters. Many parties were fitted out in 1826 for trapping on the Gila, San Francisco and Colorado Rivers. Some had licenses and some had not. Their activity contributed to the alarm with which Smith's return to California was greeted by the Mexican authorities. In October, 1826, a company of about 100 "Anglo-Americans" was reported trapping in old Sonora, New Mexico and Arizona. Doubtless some of these formed the party which divided at the Mojave settlements.

55. From this point in the transcript *Journal,* one half page and one whole page are blank, and two succeeding pages are missing. The narrative is resumed to relate events after the attack by the Mojaves.

56. Lacking the *Journal* account, we have no first-hand source of information save this sentence: "After trade and intercourse with the Indians was over, Mr. Smith and his party, in attempting to cross the river on a raft, was attacked by those Indians and completely defeated with a loss of ten men and two women (taken prisoners) the property all taken or destroyed." *Brief Sketch of Accidents, Misfortunes and Depredations Committed by Indians, etc., on the Firm of Smith, Jackson & Sublette, Indian Traders on the East & West Side of the Rocky Mountains, since July, 1826, to the Present, ——— 24th, 1829.* In record book containing copies of letters from Indian agents and others to Superintendent of Indian Affairs at St. Louis, 49-54. Kans. Hist. Soc. *Mss.* An anonymous, second-hand account was published in London in 1853 (Reprinted as *Traits of American Indian Life,* San Francisco, 1933), supposedly by Peter Skene Ogden, who knew Smith. The author affirms he had the story from Smith, but some of the details of Smith's mishaps are at variance with known facts. F. M. Kelly, Needles, California, is now, and was, forty years ago, well acquainted with the elders in the Mojave tribe. He tells me: "One charge against the Smith men was only by inference; but there seems no doubt there was considerable controversy about payment for help."

57. Eight besides Smith; after two had been left at San Bernardino, there were still seven in the party which went on to join the members of the first expedition. *Cf. Brief Sketch,* cited above. The *mayordomo* at San Bernardino reported to Ensign Santiago Argüello that Smith had eight

companions. *Argüello to Echeandia, Feb. [Oct.] 8, 1827. D.S.P., ii,* 35. Bancr. Lib.

58. The Sink of the Mojave, or Soda Lake.

59. Vanyumes. The tribe is now extinct. The Piutes (Pah-Utes, Pioches, Pautehs, etc.), called Pautch and Pauch by Smith were a numerous nation scattered in eastern California, Nevada and Utah. Both the whites and the warlike Indian tribes looked upon the Piutes as inferior natives. Smith named Beaver Dam Wash Pautch Creek because of the Piutes he met there on his first California expedition.

60. Cajon Pass between the San Gabriel and San Bernardino ranges of mountains.

61. Not the site of the present city of that name, but a point, on the Rancho of San Bernardino, southeast of the modern city and just outside what is now Redlands. The Franciscans planned to build there one of an inland chain of missions, but because of secularization it never advanced beyond the stage of an *asistencia,* or branch. This inland mission plan, overlooked by historians, is set forth by George W. Beattie in *California's Unbuilt Missions,* Los Angeles, 1930. When Smith visited the ranch there was a large warehouse there for storing products of Indian agriculture. An irrigation canal had been built, under direction of the padres, to carry water from the mountains.

62. Galbraith, a man of gigantic size and a crack shot, later settled at Los Angeles, where, it is recorded, he amused himself daily by shooting off the heads of blackbirds, at twenty paces. He was the messenger who carried Smith's letter to Father Sánchez at San Gabriel. The letter averred that when Smith left San Gabriel in the winter before he did not expect to return to California, but that when "above San Francisco" (presumably the Mojave village of that name) he encountered many Indians who at first acted in a friendly manner, and then, while he was crossing the river, attacked him, killing eight men, taking all the animals and the goods they carried, leaving the men who escaped only with the loads they carried on their backs. He notified Father Sánchez that he had killed four beeves of the mission herds, and concluded by thanking him for the benefits he received when there before. Father Sánchez gave the letter to Argüello, to be forwarded to Echeandia. *Argüello to Echeandia, Feb. [Oct.] 8, 1827. D.S.P., ii. 1821-1830,* 35. Bancr. Lib.

63. Survivors of the Mojave attack were: Thomas Daw, Toussaint Maréchal, Thomas Virgin, Isaac Galbraith, Joseph Palmer, Joseph Lapoint, Charles Swift and John Turner. *Cf.* note 46.

63a. Deleted.

64. The men left on the Appelamminy were: Harrison Rogers, Louis Pombert, James Read, Abraham LaPlant, John Gaither, Emanuel Lazarus, Peter Ranne, Martin McCoy, John Reubascan, Arthur Black and John Hanna. Pombert and Read later deserted.

65. An Appelamuny River appears on the map of Lieutenant Charles Wilkes, U.S.N., (1841), as the uppermost branch of the San Joaquin. Several entries in the *Journal* indicate that Smith's Appelamminy is the modern Stanislaus, and in the Bojorges, J., *Recuerdos sobre la historia de California*, Bancr. Lib. *Mss.*, the river on which the Americans were encamped was called Río Estanislao, because Estanislao, a truant Indian who was having trouble with the authorities, took refuge on its banks. Smith's *Journal* notes that his Appelamminy camp was seventy miles northeast of Mission San José. The Apelamenes were a San Joaquin River tribe. It is likely that Smith made no distinction between them and the Taulamnes, who dwelt on the Stanislaus. Similarly, he (or his copyist) appears to have accepted the Mokélumnes and the allied Yatchachumnes as the same tribe, spelling the name variously as Macalumbry, Mackalumbry, Machalunbry and Machallumbry.

65a. Deleted.

65b. Deleted.

66. The Mokelumne River. Smith's Macalumbry River is evidently the Calaveras.

67. Meanwhile down in Southern California the civil authorities were conducting an investigation of Smith's reappearance, upon order of Echeandia. A certain Manuel admitted having guided Smith's party from the mouth of the Cajon de Muscupiabe (Cajon Canyon) to an Indian village, Otongallavil, on the Mojave River (near the present Hesperia). Manuel said the men were nearly naked. Santiago Argüello, *alferez,* or military ensign, was ordered to demand a passport from Smith, showing that the *comandante general* permitted him to make this reappearance; if he did not have one, he should be detained. *Argüello to Echeandia.* See note 62. The *comandantes* at San Diego, Santa Barbara and San Francisco were ordered to get information "with the utmost firmness regarding the American fishermen captained by Smith, the opinion being that it is not possible to trust them further, and they must be held wherever found." *Echeandia to Comandantes, Sept. 14, 1827. D.S.P., vii,* 88

68. On May 15 and May 16, 1827, four hundred neophytes ran away from Mission San José and went to the Tulare Lake region, whence from time to time horse thieves harried the missions. Smith did not know that a few days before he left for Salt Lake, Father Durán had written an indignant letter to Ignacio Martinez, *comandante* at the *presidio* of San José, charging that "the Anglo-Americans" were urging Indians to abandon the missions and offering them protection if they returned to their villages; that these men had been in the rancherías of the Muguelemnes and Cossmines and "this inconvenience . . may be the beginning of such troubles . . . in other missions." Father Durán believed "them to be the same people who . . . have come all along the line of missions causing trouble." *Durán*

to Martinez, Archivo del Arzobispado de San Francisco, v, part i. It is not improbable that some of the American party, noting the docility of the Mission Indians, told them of the glory of the murderous Blackfeet; Harrison Rogers, for example, looked with disapproval upon this docility. It must be remembered that the early part of the nineteenth century was a period of intense religious bias, and that as a result of national rivalry Spanish practices were anathema to English-speaking people. However, in a confidential report to the Governor, Martinez said he suspected that a certain Indian, Narciso, who informed against the Americans, actually advised the runaway himself, and other Indians blamed him instead of the white men. *Martinez to Echeandia, Ibid,* 28-29.

69. William Welch, a sailor who had been discharged from an English ship at Bodega.

70. Alfred Robinson, who came to California in 1829, as clerk on the *Brookline,* found Father Durán kind, benevolent and blessed by many a wayfarer; but Auguste Duhaut-Cilly, master of the French trader *Héros,* said that in 1827 the missionary was low in spirits, worried about secret societies among the Mexican people and the secularization movement which threatened to interfere with the Franciscan plan for the spiritual and temporal welfare of the savages. Robinson, *Life in California,* New York, 1846, and Duhaut-Cilly, *Voyage autour du Monde,* Paris, 1834. Translation of Duhaut-Cilly's California material by Charles F. Carter is in Cal. Hist. Soc. *Quarterly,* no. 3, September, 1929.

71. The nearest thing to a joke in all the known writings of Jedediah Smith.

72. Ignacio Martinez.

73. Jedediah supposed the Peticutsy to be the San Joaquin River, on which the Pitcachi or Pidekati Indians resided. These Indians were reported extinct in 1877.

74. Precisely the reason.

75. Smith and his men are referred to in some official correspondence as *pescadores,* fishermen, from their custom of setting traps in streams.

76. John Rogers Cooper (called Juan *el manco*). He was born on the island of Alderney in the English channel, and in his childhood was taken to Massachusetts. He came to California at the age of twenty-one, as master of the ship *Rover.* He settled at Monterey, where he was given the Spanish names Juan Bautista.

77. Gap in the narrative.

77a. Deleted.

78. William Edward Petty Hartnell (Guillermo Arnel), an Englishman who came to California on the *John Begg* as a member of the firm of McCulloch, Hartnell & Company, agents of Begg & Company, Lima. Hartnell traded with the missions for hides and other produce. He knew

French and German, besides Spanish and his native tongue. The fact that Echeandia, who was already well acquainted with Jedediah Smith, began talking to him in Spanish, indicates that Jedediah must have acquired some knowledge of the language, probably before making his southwest expedition. Several statements in the *Journal* support this assumption.

78a. Deleted.

79. Daniel Ferguson, who hid from Smith when the party was leaving San Gabriel, and John Wilson, who was discharged near San Bernardino, but could not get permission to stay in the country. Wilson was taken back, without pay, but again discharged at Tulare Lake. He was imprisoned at Monterey; after his release he remained and married. Ferguson also remained. *Comandante* Martinez informed Echeandia that he "had a report that Juan Wilson says that they [Smith party] have been traveling eighteen months from Boston for the purpose of making maps and buying land from the natives." *Martinez to Echeandia, Archivo del Arzobispado de San Francisco,* v, part i, 28-29. Chintache Lake is Tulare Lake, King's County, California. The Wilkes map shows the name of the lake as Smith spelled it—Chintache. Bonneville shows it as Chataqui, Burr as Chentache. The lakes of the Tulare Valley, surrounded by tules, or bulrushes, were a secure hiding-place for gentile Indians and neophyte deserters, who raided the livestock of the Mission Indians, ran horses to exhaustion, then killed and ate them.

80. William G. Dana, master of the ship *Waverly,* came to California in 1826. He married a girl of the country and begat twenty-one children. He engaged in trading, agriculture, stock-raising and soap-making.

80a. Deleted.

80b. Deleted.

81. Echeandia was *comandante general.*

81a. Deleted.

81b. Deleted.

81c. Deleted.

81d. Deleted.

81e. Deleted.

82. Nothing further is known of such a company of Americans, if there was one, although *Humphrey to Gwin,* 1858, Bancr. Lib. *Mss.,* says that "Richard Campbell of Santa Fé" came with pack-mules from New Orleans to San Diego in 1827.

83. Captain John Bradshaw of the ship *Franklin* traded along the coast of California. In later years the Mexican authorities accused him of smuggling and other offenses, including "insolence to the Governor." He was wounded when his ship was fired upon as he fled to escape arrest.

83a. Deleted.

83b. Deleted.

83c. Deleted.

83d. Deleted.

84. "I, Juan Bautista R. Cooper, a citizen of the United States of America, residing in this country, declare that I have been designated as their representative by José Stil [Joseph Steele], captain of the barkantine *Harbinger,* and Tomás B. Parc [Thomas B. Park], supercargo of said barkantine, and by Henry Pease, mate of the schooner *Hesper.* [The reference to the *Hesper* in the bond is incomplete; thus: "Henry Pease 2° commandante de la Frigate (*sic*) Hesper Benjamin clark". When the *Hesper* left Nantucket in August, 1826, Pease was captain. Benjamin Clark was captain of various whaling vessels of the same ownership in later years.]

"Therefore, in the name of the Government of the United States, I provisionally guarantee the good conduct and behavior of Captain Smith in whatever concerns his return to the settlement called 'the deposit', and for the furnishing of such equipment as shall be necessary, I hereby bind myself and all property in my possession or to be received, to be responsible to the Republics of Mexico and of the United States, and to anyone who may be concerned, for each and every one of the following articles.

"1. That Captain Smith and his men are citizens of the United States, honest and faithful to our Government, and as such they should be considered friends and bound by the same agreement as exists between the two nations.

2. That for this reason such help as shall be absolutely necessary for their return, such as arms, ammunition, horses, provisions, etc., shall be furnished them at their just prices, the same being for the purpose of protecting them and their property until their safe return to the settlement called the Salt Lake deposit, following the road from Mission San José by way of Carquinez Straits and Bodega.

"3. That under no condition will he delay on the way a longer time than is necessary, and that having reached his destination, he will make no hostile excursion, and will make no trip toward the coast or in the region of his establishment south of the 42nd parallel not authorized by his Government in accord with the latest treaties, unless he has a legal passport expressly [permitting it] from one or the other of the aforesaid Governments.

"4. Four copies of this direct guarantee to the Governor of California shall be drawn—one to be filed with the Governor, one to be retained by Captain Smith, one to be sent to Mexico, and the other to be kept in my possession. In witness whereof, we have set our hands in Monterey, November 12, 1827. (Signed) Juan Bautista R. Cooper.

"I acknowledge this Bond. Jedediah S. Smith." Vallejo, M. G., Bond in *Documentos para la historia de California,* xxix, 171. Bancr. Lib. *Mss.*

85. Smith signed another bond, probably for the protection of Captain Cooper: "I, Jed[a] S. Smith, of Green Township in the State of Ohio, do hereby bind myself, my heirs, executors and principals in the sum of thirty thousand dollars for the faithful performance of a certain Bond, given to the Mexican Government; dated at Monterey, November 15, 1827." In Sublette *Papers,* Miss. Hist. Soc. *Mss.*

86. "In consideration of the foregoing security, I hereby grant free and safe passport to Captain Smith, in order that, accompanied by the seventeen men he brought under his command, he may return to his settlement, each man carrying his own fusil or gun; a total of seventy-five pounds of powder and one hundred and twenty-five pounds of lead, five loads of clothing, and other goods, six loads of provisions, two loads of merchandise for the Indians, one load of tobacco, and other loads comprising the equipment he brought; a total of one hundred mules and one hundred and fifty horses. José M.[a] de Echeandia." *Vallejo Documentos,* xxix, 173.

87. Captain Luis Antonio Argüello. He was born at San Francisco *presidio* in 1784. He entered military service in 1799 and rose until he was made captain in 1818. Four years later he was chosen acting Governor of California. He resigned in 1825. Echeandia relieved him of his military command in 1828, on the ground of Argüello's reputed ill health.

88. "They had not" and rest of this sentence in *Journal* written in another hand, very like Smith's own.

89. Henry Virmond.

90. Frederick William Beechey, commander of the *Blossom,* was a famed explorer, who had coöperated with Franklin and Parry in their polar expeditions. At the time Smith met him he was making observations in and about the San Francisco Bay region.

91. *Fulham.*

92. William R. Garner, who had deserted an English whaler at Santa Barbara.

92a. Deleted.

93. Rancho San Pablo, owned by Don Francisco Castro.

94. Father Luis Gil y Taboada.

95. A holyday, rather, in honor of the Immaculate Conception of the Blessed Virgin Mary.

96. The words of the Mass were in Latin; the sermon in the Spanish and the native languages.

96a. Deleted.

97. The *Brief Sketch* makes the charge that "the governor had instructed the Muchaba Indians not to let any more Americans pass through the country on any conditions whatever; to this advice Mr. Smith leaves the entire cause of his defeat." No contrary testimony of provocation has

come to light, as in the case of the later Umpqua disaster. It is not improbable that the Mexican civil authorities sent presents to the Mojaves and asked them not to let any more aliens pass through their territory; but it is very doubtful they were "instructed to kill all Americans." The Mojaves surely were taking no orders from the California authorities, else there would have been no stolen horses and truant mission Indians among them. Moreover, the presence of two Indian women among the whites must be considered an incentive for attack.

98. Captain William A. Richardson disembarked from a whaler at San Francisco in 1822, and was permitted to remain in California on condition that he teach navigation and carpentry. He married a daughter of Ignacio Martinez in 1825. He built a launch, which was used for hire and for collecting produce for the pueblos. He served as pilot in S. F. Bay and in 1837 was captain of the port.

98a. Deleted.

99. The ranch of Francisco Castro.

100. The Rogers *Diary* in Miss. Hist. Soc. *Mss.* lists Leland as Richard Layla[n]. The name is so indistinct it has been mistaken for Taylor. The sixth letter appears to have been torn away.

101. Captain John Burton, who said he had been master of the ship *Juan Battey,* lost at San Diego. He came from Provincetown, Mass., settled in California and married there.

102. I. e., a sheep farm called San Lorenzo, belonging to the mission.

103. Sacramento River. Smith evidently considered Carquinez Strait the mouth of the Buenaventura. Until he returned from California it was commonly believed in the States that the Buenaventura River flowed westward from the Rockies.

103a. Deleted.

104. Guadalupe River.

104a. Deleted.

104b. Deleted.

104c. Deleted.

104d. Deleted.

105. Jedediah is somewhat unjust to the nation which produced such great explorers of the West as Anza, Garcés and Escalante, though perhaps he was unaware of their achievements. The party of discovery to which Smith refers was doubtless that of Captain Luis Argüello and Father Blas Ordaz, who, in 1821 went up the Sacramento approximately to the site of the present town of Chico, turned west and came back by way of Russian River Valley. In 1817 Argüello, Father Durán and Father Ramón Abella went up the river Sacramento about to the mouth of the American fork, returned and ascended the San Joaquin as far as Calaveras River. The padres took note of Indian villages which might be brought under

Christianizing influence of the missions. On the 1817 journey, at the end of a week, Father Durán recorded in his diary: "At about five o'clock, looking through a gap in a grove of the river bank, we discerned the famous Sierra Nevada. The white part of this Sierra seemed to be all snow, although, as they say, it also has a species of white rock which looks like snow." Academy of Pac. Coast Hist. *Pubs.,* ii, no. 1, 339. Several days later the diarist observed: "Once the pass in the Sierra is discovered, which the said end [of the mountain chain] seems to offer, we would be able to ascertain the truth of what the Indians have told us for some years past, that on the other side of the Sierra Nevada there are people like our soldiers. We have never been able to clear up the matter and know whether they are Spanish from New Mexico, or English from the Columbia, or Russians from La Bodega." *Ibid.,* 343.

106. Echeandia had arranged for an escort of soldiers to accompany the Smith men and make sure they left Mexican territory. When Smith ignored the Governor's arrangements, Echeandia wrote to the *comandante* at San Francisco a letter regarding "the bad behavior and abuses of the American, Smith." The letter was destroyed in the San Francisco fire of 1906. The Mexican Government complained to the American Government about Smith's invasion, and the American Government replied that Smith affirmed he was treated harshly in California. *House Executive Documents, 25th Cong. 2nd Sess., Doc. 351,* 246-248.

107. Not the Buenaventura, which is the modern Sacramento, but the Old River branch of the San Joaquin, sometimes then called the Pescador.

108. San Joaquin River.

109. Lone Tree Creek or French Camp Slough.

110. Louis Pombert, a Canadian. He settled in California and married a member of the Pico family.

111. Mokelumne River.

111a. Deleted.

112. The main Calaveras River formerly ran through Mormon Slough channel at the south edge of what is now Stockton.

113. Sentence beginning "So I returned" written in another hand.

114. Read had been troublesome at San Gabriel, and was given a flogging by Smith. There is no record of him after he deserted the Smith party, although in 1860 a certain "Bill Reed" was reported at Mono Diggings, near Mono Lake, claiming to have been there "in 1825" with Smith's party, which spent a week successfully prospecting for gold. The only mention of gold in Jedediah's *Journal* probably referred to the metal in the sense of wealth, and the best informed living member of the Smith family, who frequently talked to Jedediah's brothers about "Uncle Jed's" adventures, heard nothing of a gold discovery.

114a. Deleted.

115. American River.

116. American River. Reclamation has changed conditions in this area.

116a. Deleted.

117. Bear River.

118. Feather River.

118a. Deleted.

119. Yuba River. Deposit of mining debris has changed the location and appearance of this stream.

120. Yuba.

120a. Deleted.

121. Feather River.

122. Honcut Creek.

122a. Deleted.

123. Gap.

123a. Deleted.

124. The bars have been worked over for gold and now are a series of dredger tailing piles.

125. Butte Creek.

126. Sacramento.

127. Chico Creek.

128. Deer Creek.

129. Stony Creek.

130. Toomes Creek.

131. Mill Creek.

131a. Deleted.

131b. Deleted.

131c. Deleted.

131d. Deleted.

131e. Deleted.

132. Smith had crossed the divide into Trinity National Forest, Trinity County, California. He crossed the Buenaventura (Sacramento) south of the modern Red Bluff.

132a. Deleted.

132b. Deleted.

133. Smith's River is evidently the Trinity, and not the stream shown on modern maps as Smith River. Smith and others gave the name Clamouth to the river now called the Rogue. Harrison Rogers, whose diary parallels Smith's beginning with the entry of May 10, 1828, called the Trinity by the name Indian Scalp River.

133a. Deleted.

133b. Deleted.

133c. Deleted.
133d. Deleted.
133e. Deleted.
133f. Deleted.
134. The Klamath River, into which the Trinity flows. Smith was now between Pine Creek and the west bank of the Trinity, the northern part of the present Hoopa Valley Indian Reservation.
134a. Deleted.
134b. Deleted.
134c. Deleted.
134d. Deleted.
135. The Rogers diary parallels the Smith *Journal* from this point.
135a. Deleted.
136. Smith evidently supposed the Klamath a continuation of the Trinity.
136a. Deleted.
136b. Deleted.
137. Smith crossed the Klamath at a point northeast of the present town of Klamath, and south of the Del Norte-Humboldt County line.
138. Jedediah did not think it worthy of mention that on this day he "got kicked by a mule and hurt pretty bad," as Rogers records.
138a. Deleted.
139. Peter Ranne, the Negro.
140. In the Rogers *Ms.* this word looks like "kitten," though perhaps he wrote "kittle."
141. Wilson Creek.
141a. Deleted.
141b. Deleted.
141c. Deleted.
141d. Deleted.
141e. Deleted.
141f. Deleted.
141g. Deleted.
142. Camas, or cammass.
143. Lake Earl, Del Norte County, California.
144. Smith River; one of the two streams now named for Jedediah.
144a. Deleted.
145. Chetco River, Curry County, Oregon.
145a. Deleted.
145b. Deleted.
145c. Deleted.
145d. Deleted.
145e. Deleted.

146. Rogue River.

147. Garrison Lake. Identification of points between the Rogue and Coquille is difficult; Smith and Rogers differ in their record of mileage.

148. Sixes River.

149. Floras Lake.

150. Coquille River.

150a. Deleted.

151. The Rogers diary continues until July 13.

152. Dr. John McLoughlin was a Canadian, born of an Irish father and a Scotch mother. He was well over six feet in height, and impressive of mind as well as of body. As chief factor, he ruled an empire from Fort Vancouver, except when his superior officer, George Simpson, was present.

153. Letter in Hudson's Bay Company *Archives,* Fort Vancouver Correspondence Book, 1828-1829, No. 927, folios 28d, 29. This letter and other material from the H.B.C. *Archives* are reproduced by permission of the Governor and Committee of the Company, at London.

154. *Ibid.*

155. Actually there are two journals of this expedition, both apparently kept by McLeod. One contains full details of the expedition from September 6, 1828, to October 17 of the same year; the other is complete from October 17 to and including December 10, but the account of events from September 6 to October 17 is summarized from memory. The second journal is called the *Journal of the Columbia Southern Expedition.* These documents are numbered 968 and 969 in the H.B.C. *Archives.*

156. Falls of the Willamette.

157. Variant spelling of Champoeg (Champoick, Champooing, Shampooick, etc.) The site of Champoeg is about midway between the present Newberg and Butteville, Oregon.

158. Thomas McKay was the son of Alexander McKay, who was lost in the attack by Indians on the Astorian ship *Tonquin.* Thomas McKay was half Scotch, half Indian; the idol of the half-breeds of the Northwest, who admired his bravery and marksmanship. Michel LaFramboise came to Oregon on the *Tonquin.* When Astoria was surrendered to the British he joined the Northwest Company, later he entered the service of the H.B.C.

159. Near the junction of Calapooya Creek and Umpqua River.

160. It was the custom of the H.B.C. to give their men starting on an expedition a feast or *regale.*

161. McLeod supplementary journal.

162. Wild horses.

163. Santiam.

164. Mill Creek? Chembukte resembles Cheméketa, the old name of Salem.

165. The present Calapooya River.

166. Nomtomba (Lung-tum-ler, Lumtumbuff, Long-tongue-buff, etc.) has become Long Tom!

167. Elk Mountain and Elk Creek.

168. Cayuse?

169. The Verveau was probably at the mouth of Mill Creek, west of the modern Scottsburg. An Indian named Verveau was in the employ of McLeod. The diary of John Work's Umpqua expedition uses the word Vervor. Ore. Hist. Soc. *Quarterly,* v. xxiv.

170. Logan, Scott, O'Hara and Bell never reappeared among their fellows. The casualty chart accompanying the *Brief Sketch* says, "The fate of these men is not known, but the conclusion is hardly doubtful."

171. There is no record of finding remains of the other four men of Smith's company. It is possible they were carried off as prisoners and later put to death.

172. Tahkenitch Creek.

173. Siltcoos River.

174. Siuslaw.

175. Killamoux or Tillamooks.

176. Ten-Mile Creek?

177. The present Smith River, Douglas County, Oregon. The Burr map shows it as Defeat River.

178. Rigging.

179. Joe is one of several Indians mentioned in documents of early Oregon. He, Charles, Verveau and others were often in the employ of the H.B.C. traders. Joe was reputed to have half a dozen wives in domicile and an equal number without.

180. Probably a variant of Nomtomba, or Long Tom River. McLeod, Smith and others of that day frequently spelled one word in a variety of ways. Much of the Smith *Journal* may have been dictated on different days; that would account for the varied spelling.

181. H.B.C. *Archives.* Simpson (George) Correspondence Book (Outward), 1828-829, no. 950, 23-29.

182. *Ibid.,* 30, 31.

183. H.B.C. *Archives.* Simpson (George) Reports Box 4, 1828, no. 924, 81-91, 94-96.

184. The Rogers *Ms.* shows that there was no imprisonment. Though technically detained, the visitors had the freedom of the place, went hunting and took employment. Smith and Rogers were constant guests at the missionaries' table.

185. Smith says he did not go to San Gabriel, but stopped at San Bernardino. He was made prisoner at San José.

186. More likely ten shillings.

187. Dr. McLoughlin in *The McLoughlin Document,* Oregon Pioneer Association *Transactions,* 1880, 47, 48, says, "But to gratify their passion for women, the men neglected to follow the order, allowed the Indians to come into camp," etc.

188. H.B.C. *Archives.* General Correspondence Book, no. 622.

189. Simpson was born in Scotland in 1792. Thirty years later he became Governor of the territories of the H.B.C. west of the Rocky Mountains. In 1826 he was made supreme Governor of all the H.B.C. territories in North America. He was knighted in 1839. He is represented as having been very much worried over the expansionist policy of the American Government.

190. Reprint from *St. Louis Beacon* in *Daily National Intelligencer,* Washington, D. C., November 1, 1830.

191. Anon., *Jedediah Strong Smith, Illinois Monthly Magazine,* 396.

192. Letter in *Archives* of City of Mexico, *Secretaria de Relaciones Exteriores, serie segundo, caja* 1830-1834.

193. Copy of letter in Peter Smith *Papers,* among papers of late Walter R. Bacon, Los Angeles.

194. Ralph Smith, oldest brother of Jedediah, was born in 1794, probably in Connecticut. He learned the millwright trade under an uncle. After the Smith family left New England, he lived successively in New York, Ohio, Indiana and Michigan. In his later years he suffered from the effects of an accident, which confined him to bed for three winters. Ever afterward he walked with crutches. He died in 1867 in Eaton County, Michigan. His wife, Louisa Simons, died in 1874, near Dexter, Iowa.

195. Austin Smith, brother of Jedediah, was born in 1808, probably at Bainbridge (Jericho), N. Y. He accompanied Jedediah on the journey over the Santa Fé trail in 1831. For a time he was engaged in business at St. Louis. He died at an early age, suddenly, while on a business trip in Illinois.

196. Peter Smith, brother of Jedediah, was born in 1810. He was apprenticed to the tinner's trade, but did not continue in that work. In 1831 he accompanied Jedediah on the Santa Fé trail, and from Santa Fé went into California with a company of Jackson, Waldo & Young. He soon returned to Santa Fé, and after staying in New Mexico several years, went to Ashtabula, Ohio, where in 1836 he married Juline E. Babcock. Mr. and Mrs. Smith moved to Mt. Pleasant, Iowa, in 1841. Peter Smith went to Guanajuato, Mexico, in 1843, to meet Samuel Parkman, Jedediah's amanuensis, with whom Peter had been engaged in business at Santa Fé under the firm name of Parkman & Smith. In 1850 he and his wife and children joined a caravan going to California. This was one of several cross country journeys this family made. Peter Smith accumulated considerable gold in California. He settled in Council Bluffs, Iowa, where he obtained a street-

sprinkling contract. Later the family moved to Grand Island, Nebraska, and there Peter Smith died.

197. Ira Gilbert Smith, brother of Jedediah. He was born in 1811, probably at Bainbridge. After the death of Jedediah Smith, Ira left school at Jacksonville, Illinois, and went to St. Louis. In 1832 he displaced General Ashley as administrator of Jedediah's estate, and soon afterward filed an additional inventory of assets, including the Smith *Journal,* from which I quote. He went with a caravan to Santa Fé, and there for a while conducted a store. He married in Iowa in 1836. In 1849 he went with a wagon train to the California gold fields, and afterward served as constable at Sacramento. Upon his return from California he settled at Greenville, Illinois.

198. Benjamin Greene Paddock Smith, brother of Jedediah. He was born in 1813, at North East, Erie County, Pennsylvania. He was named after a Methodist clergyman then much admired. Upon leaving school he was apprenticed to a saddler, but he did not remain at the trade; he opened a mercantile business with his brother Austin, in St. Louis, and later with his brother-in-law, James Young. (He married Margaretta Nye Young in 1834). After leaving St. Louis he farmed for several years in Morgan County, Illinois, and later on a piece of land on the Missouri River, close to St. Louis. Thence he moved to Greenville. where he remained until letters from his brothers Peter and Ira attracted him to California. He accompanied a wagon train, which met so many hazards that the argonauts reached California minus wagons, stock and all other property save their rifles and horses. At Sacramento Paddock became deputy under his brother, Constable Ira G. Smith. Later he went into business in Greenville. Again he started for California overland but Mrs. Smith was taken ill at Oskaloosa, Iowa, and the journey abandoned. Eventually he settled in Texas, where he died.

199. Nelson Jones Smith, youngest member of the Smith family, was born in Erie County, Pennsylvania, in 1814. In 1834 he married Sarah Ann Crandall. N. J. Smith was a merchant; he conducted a jewelry store in Ashtabula, Ohio, whence he moved to Mt. Pleasant, Iowa. Later in Oskaloosa, Iowa, he conducted a hardware store. For a time he engaged in mining at Silver Cliff, Colorado. The burning of his home in Oskaloosa probably destroyed much data about Jedediah Smith. His grandson, Guy S. Carleton, recalls there were numerous Indian and pioneer relics in the house.

200. Sally Smith, eldest child of Jedediah Smith, Sr., was born in Connecticut in 1791. She was married first to a man named Shiffer, and later to William Jones. She lived most of her life in Ashtabula, Ohio.

201. Betsey Smith was born in 1796, probably in Connecticut. She was married to Edward R. Davis in 1818. She died at Wayland, Iowa, in 1851.

202. Eunice Smith was born in 1797. She was married to Solomon A. Simons, son of Dr. T. G. V. Simons, and later to a man named Beers. Eunice Simons was known to the Indians as "The Great White Medicine Woman."

203. Jedediah sent to Ralph Smith a considerable amount of money to give or lend to relatives and friends, besides $1500 for the purchase, in Jedediah's name, of a farm owned by a certain Major Tiller (Tyler?) in Wayne County, Ohio. In addition, as a result of his activities in the mountains and the Santa Fé trade, there were several notes outstanding and due him, including one-third part of $15,132.22, from Thomas Fitzpatrick, Milton G. Sublette, Henry Fraeb, James Bridger and J. B. Jarvis (Gervais), and the whole of $2850 for goods sent to Santa Fé and purchased by the same partnership.

Index